Apple Watch Manual for Beginners

The Perfect Apple Watch Guide for Seniors, Beginners, & New Apple Watch Users

Covers all Apple Watch models

Joe Malacina

An Infinity Guides book

InfinityGuides.com

Apple Watch Manual for Beginners

Copyright 2024 by No Limit Enterprises, Inc., All Rights Reserved

Published by No Limit Enterprises, Inc., Chicago, IL

No part of this publication may be reproduced, copied, stored in a retrieval system, or transmitted in any form or by any means, electronic, digitally, mechanically, photocopy, scanning, or otherwise except as permitted under Section 107 or 108 of the 1976 United States Copyright Act, without the prior express written permission of No Limit Enterprises, Inc., or the author. No patent liability is assumed with respect to the use of information contained within this publication.

Limit of Liability & Disclaimer of Warranty: While the author and No Limit Enterprises, Inc. have used their best efforts in preparing this book, they make no representations or warranties with respect to the accuracy or completeness of the contents of this book and specifically disclaim any implied warranties of merchantability or fitness for a particular purpose. No warranty may be created or extended by sales representatives or written sales materials. The advice, strategies, and recommendations contained in this publication may not be suitable for your specific situation, and you should seek consul with a professional when appropriate. Neither No Limit Enterprises, Inc. nor the author shall not be liable for any loss of profit or any other commercial damages including but not limited to: consequential, incidental, special, or other damages.

The views and/or opinions expressed in this book are not necessarily shared by No Limit Enterprises, Inc. Furthermore, this publication often cites websites within the text that are recommended by the author. No Limit Enterprises, Inc. does not necessarily endorse the information any of these websites may provide, nor does the author.

Some content that appears in print may not be available in electronic books and vice versa.

Trademarks & Acknowledgements

Every effort has been made by the publisher and the author to show known trademarks as capitalized in this text. No Limit Enterprises, Inc. cannot attest to the accuracy of this information, and use of a term in this text should not be regarded as affecting the validity of any trademark or copyrighted term. Apple Watch® is a registered trademark of Apple, Inc. All other trademarks are the property of their respective owners. The publisher and the author are not associated with any product or vendor mentioned in this book except for Infinity Guides and The Digitize Center. *Apple Watch® Manual for Beginners* is an independent publication and has not been authorized, sponsored, or otherwise approved by Apple, Inc. Other registered trademarks of Apple, Inc. referenced in this book include: iPhone®, iPad®, Mac®, AirDrop®, and others.

Warning & Disclaimer

Every effort has been made to ensure this text is as accurate and complete as possible. This text does not cover every single aspect, nor was it intended to do so. Instead, the text is meant to be a building block for successful learning of the subject matter of this text. No warranty whatsoever is implied. The author and No Limit Enterprises, Inc. shall have neither liability nor responsibility to any person or entity with respect to any losses or damages arising from the information contained in this text.

Contact the Publisher

To contact No Limit Enterprises, Inc. or the author for sales, marketing material, or any commercial purpose, please visit www.nolimitcorp.com, or email info@nolimitcorp.com.

Apple Watch® Manual for Beginners

Publisher: No Limit Enterprises, Inc.

Author: Joe Malacina

About the Author

Joe Malacina is the founder of *InfinityGuides.com*, a beginner's help website that offers books, DVDs, and online courses to help people learn how to use technology. Since 2013, *InfinityGuides.com* has taught over 100,000 people how to use their devices and how to use social media. Prior to authoring numerous how-to books, Malacina operated two popular tech blogs with a combined audience of over one million people that focused on the most popular smartphones and tablets. He has also been a guest speaker at technology and adult education conferences around the United States. His approach for teaching technology devices has focused on stressing the basics of building technological intuition, thus allowing users to become experts at using their devices in a relatively short amount of time. Following the success of his blogs and seminars, Malacina authored his first book, the *iPhone Manual for Beginners (1st Edition),* which was published in 2017 and reached the *InfinityGuides.com's* Best-Seller list within the first week and remained there for over six months. His published works have earned praise from notable tech aficionados, adult and senior organizations, and readers alike, and can be found in households across the United States, Canada, Australia, and the United Kingdom.

Malacina holds an MBA in Finance from the University of Illinois at Chicago and holds a Bachelor of Science in Mechanical Engineering from the same university. He has been profiled and interviewed on numerous radio stations, podcasts, and blogs and is often on a speaking tour in the United States. He lives and works in Chicago, IL, United States.

www.joemalacina.com

Also by Joe Malacina

iPhone Manual for Beginners: The Perfect iPhone Guide for Seniors, Beginners, & First-time iPhone Users

iPad Manual for Beginners: The Perfect iPad Guide for Beginners, Seniors, & First-time iPad Users

Kindle Manual for Beginners: The Perfect Kindle Guide for Beginners, Seniors, & First-time Kindle Users

Fire HD Manual for Beginners: The Complete Guide to using the Fire HD for Beginners, Seniors, & First-time Fire Tablet Users

Galaxy S8 Manual for Beginners: The Perfect Galaxy S8 Guide for Beginners, Seniors, & First-time Galaxy Users

Table of Contents

Introduction 11

Chapter 1 – About the Apple Watch 13

    Key Terms 13

    Different Apple Watch Models & watchOS Explained 14

    Cellular vs. Wi-Fi Only Models 14

    Your iPhone is Essential 14

Chapter 2 – Apple Watch Layout 15

    Turning your Apple Watch On and Off 16

    Waking your Apple Watch and Putting it to Sleep 17

    Charging your Apple Watch 17

Chapter 3 – Getting Started 19

    First-time Setup – Pairing your Apple Watch 19

Chapter 4 – Update your watchOS Now 27

Chapter 5 – Navigating your Apple Watch 29

    Digital Crown 29

    Side Button 31

    Faces 31

Chapter 6 – Faces 33

    The Watch App on your iPhone 33

    Adding Faces to Apple Watch 34

        Face Gallery 35

        Adding a Face to your Apple Watch 37

    Complications 41

    Editing or Deleting a Face that has been Added to your Apple Watch ... 45

Switching Between Faces on Apple Watches 45

Adding or Editing Faces Directly on the Apple Watch 46

Chapter 7 – The Control Center 49

Accessing the Control Center 49

Control Center Functions 50

Chapter 8 – Notifications 55

The Notification Center 55

Chapter 9 – Text Messaging 57

Sending Text Messages and the Messages App 57

Chapter 10 – Phone Calls 61

Receiving a Call 61

Making a Call 61

Call Functions 62

Tips for Using Phone Calls on your Apple Watch 63

Chapter 11 – Activity App 65

Using the Activity App 65

Chapter 12 – Fitness & Exercise 67

Workout App 67

Chapter 13 – Health Features 69

Mindfulness App 69

   State of Mind 69

   Reflect 70

   Breathe 70

Sleep App 70

ECG App 73

Heart App 75

Noise App 78

Cycle Tracking 80

Vitals App 81

Health App 81

   Viewing your Health Summary 81

Viewing your Heart Data 82

Viewing your Sleep Data 83

Viewing your Activity & Mobility Data 83

Chapter 14 – Safety Features 85

   Medical ID 85

   Contact Emergency Services 86

   Fall Detection 87

   Siren 88

   Crash Detection 88

Chapter 15 – More Useful Apps & Features 89

   Mail App 89

   Timer App 89

   Stopwatch App 90

   Weather App 91

   Compass App 91

Music App 92

Now Playing App 92

Downloading New Apps 94

Chapter 16 – Apple Pay 95

Setting Up Apple Pay 95

Using Apple Pay 97

Chapter 17 – Siri 99

Accessing Siri 99

Using Siri 99

Chapter 18 – Tips & Tricks 101

Background Apps & the App Switcher 101

Low Power Mode 102

Smart Stack 102

The Double Tap Gesture 103

Using the Camera App 103

Unlock your Apple Watch with your iPhone 105

Quickly Set a Photo as an Apple Watch Face 105

View your Apps Screen as a List 106

Chapter 19 – Conclusion & More Resources 107

More Resources – Infinity Guides 107

More Resources – The Digitize Center 108

Introduction

Congratulations! So, you have decided to take the first step, in fact the only step needed to learn how to use your Apple Watch. Maybe you do not even have an Apple Watch yet, and just want to see how it works before you decide whether to buy one. Either way, this book will teach you everything you need to know on using your device. I wanted to take this time to tell you how this book is going to be the only thing you will ever need to learn your Apple Watch. You see, this book was written from the perspective of a beginner. In other words, if you have never used a "smartwatch" in your life, that will be no detriment when reading this book. That is the big difference between this book and other competitors available. Many authors fail to realize that even today, many people are buying their first "smartwatch," and need to be shown from the ground up the basics of using and navigating their device. So that is what this book sets out to accomplish. I will teach you not only how to do specific functions on your Apple Watch, but I will teach you the building blocks of using any "smart" device. When you are finished reading this book, not only will you be a pro using your Apple Watch; you will be able to pick up any modern mobile device and have a general understanding of how it works and how to accomplish tasks.

This book is structured so that basic concepts I teach you in the beginning chapters will be used in later chapters. Therefore, I highly recommend reading the first few chapters instead of skipping ahead to exactly what you want to learn. You may miss out on essential tidbits of information that I will not cover in detail in later chapters.

Lastly, you may be wondering if this book is suitable for you. Will it answer all the questions you have? Will you understand the material? I can assure you that the information addressed in this book is derived directly from the input of several thousand Apple Watch users who have had the same questions as you. I have been teaching people how to use their mobile devices since 2013 and have taught well over 500,000 people of all ages and backgrounds how to use their devices. I have run a blog and numerous websites where I have received over 10,000 e-mails with questions on how to do this and how to do that. I know what the most common questions are, and where the most confusion lies. Take solace in the fact that this book will address your questions with a step-by-step approach, while building your technological intuition. When you are done, you will not even need to memorize the

Introduction |

steps to perform a function, you will have the intuition and knowledge to figure it out quickly. That is the core of what this book will teach you.

I welcome your feedback on this book, and you can always reach me on Twitter @JoeMalacina or on Facebook at Facebook.com/JoeMalacina1.

On a final note, Apple users, particularly iPhone and Apple Watch users, are notoriously loyal to their devices. After reading this book, you may start to see exactly why. The Apple Watch's interface and navigation are seamless and soon you will find yourself liking the way it does things. Soon after that, you will notice that many other Apple devices work in very similar ways and may find yourself wanting to explore other Apple devices such as an iPhone or iPad. If you ever find yourself wondering how these devices work, I urge you to check out www.infinityguides.com. There you can find beginners' books, manuals, and online courses on many Apple devices.

On that introduction, let us get started.

Chapter 1 – About the Apple Watch

You may now have an Apple Watch, and you are ready to start using it. So, what is an Apple Watch exactly? Your Apple Watch is classified as a smartwatch, which means that it can do most things a regular watch can do, plus additional features. With your Apple Watch, you can view the time, access notifications, send and receive messages, check your email, use apps, and access other useful features. You can do all this with a smartwatch that utilizes a touch screen. Before I show you how to do all these tasks, let us familiarize ourselves with some key terms that I will use often in this book. These terms are important for you to remember, as you will encounter them often.

Key Terms

Apps – Apps are programs on your Apple Watch and iPhone that can do tasks. Your Apple Watch comes with many apps already installed, and you can download many more through the App Store on your iPhone. Nearly every aspect within the Apple Watch is part of an app, which you will see later. Apps appear as circle icons on your Apple Watch in the Apps Screen.

Faces & Home Face – Throughout this book, you will see the term faces and home face used often. Your home face is the main screen of your Apple Watch that shows you the time and other information. This face can be changed and customized to your liking, and you can save multiple different faces that you can switch to at any time.

Bands – The Apple Watch connects to your wrist using a band. If you have not connected a band to the watch, you can do so by sliding the band into place until it is secure and "clicks". To remove a band for any reason, press and hold the small release button on the back of the Apple Watch while sliding the band out of its holster.

iPhone – In order to use your Apple Watch with all its features, you must have an iPhone. Keep your iPhone handy while reading this book as we will be using it often to get your Apple Watch completely set up.

Chapter 1 | About the Apple Watch

Different Apple Watch Models & watchOS Explained

There are many different Apple Watch models available in the marketplace, and it is important to understand the main differences between them. Generally, most Apple Watch models work almost entirely in the same manner. For instance, performing most tasks and functions on an Apple Watch Series 10 can be done in the exact same manner on an Apple Watch Series 7. The reason this is important is so you understand that all Apple Watch models operate nearly exactly the same, depending upon which software they are using. So, if one friend has an Apple Watch Series 10, and your other friend has an Apple Watch SE, they will operate in almost the exact same way in every single aspect so long as both Apple Watches are using the same software version. The main differences between Apple Watches are some feature differences and the hardware inside the smartwatch. All in all, remember this: No matter which Apple Watch you have, it will work nearly the same as every other Apple Watch that is running the same software. Any major differences that you might see on your Apple Watch will be explained thoroughly in this book.

Let us now explain the software running on Apple Watches. The software your Apple Watch is running is called its watchOS. watchOS stands for Watch Operating System, and it governs how everything in your Apple Watch works. watchOS is labeled by a number, and the higher the number, the newer the software. For instance, the Apple Watch Series 10 and Ultra 2 come with watchOS 11. The Apple Watch Series 9 originally came with watchOS 10. At all times you should be using the newest version of watchOS available for your Apple Watch. I will show you how to update the software in Chapter 4.

Cellular vs. Wi-Fi Only Models

Your Apple Watch can be a cellular model or a Wi-Fi only model. A cellular model can access all network features without having your iPhone nearby whereas Wi-Fi only models require your iPhone to be nearby. Cellular models connect to a cellular network to access data when your iPhone is not nearby or available.

Your iPhone is Essential

Your iPhone is an essential partner to your Apple Watch, and you will be using it frequently to help customize your watch. For many customizations on your Apple Watch, it is more practical to do it on your iPhone rather than directly on your watch. In this book, we will always cover the most practical way to perform a task.

Chapter 2 | Apple Watch Layout

Chapter 2 – Apple Watch Layout

Now we enter the "instruction manual" portion of this book, and we start with the Apple Watch's layout. Shown in Figure 2.1 is the layout of an Apple Watch Series 10 or similar model.

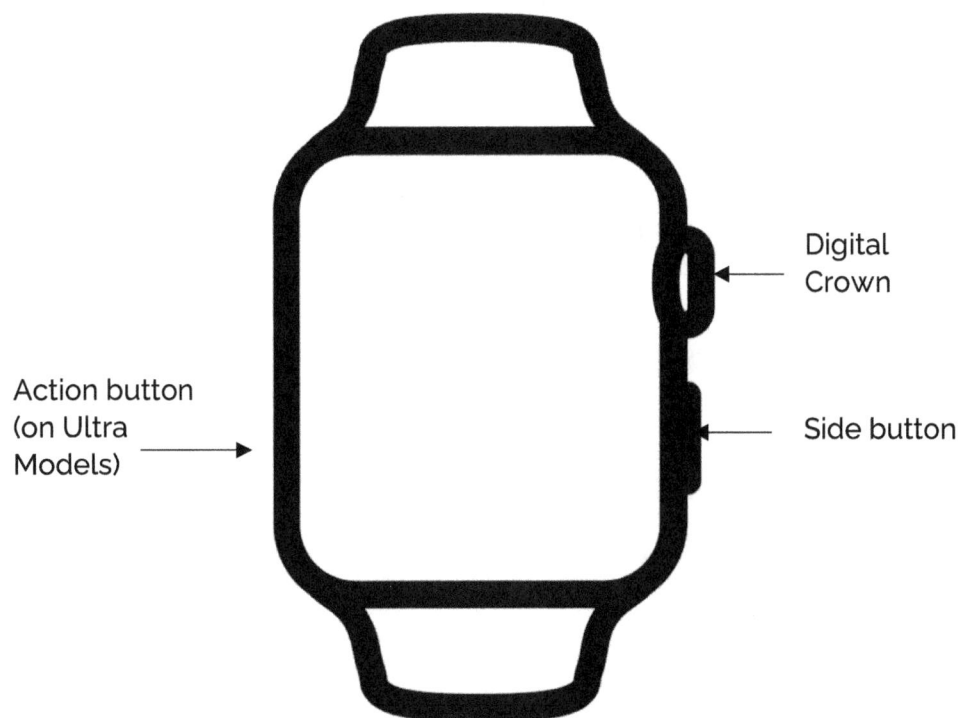

Figure 2.1 – Layout of an Apple Watch

As you can see in Figure 2.1, there are only two or three buttons on modern Apple Watches. Some specific models may have additional buttons on the left side of the device; however, you will not need them to perform most functions on your Apple Watch especially after you update its software.

- **Digital Crown** – The Digital Crown is a button you can rotate and press in. Pressing this button will return you back one screen or return you to your

Chapter 2 | Apple Watch Layout

home face. Rotating the Digital Crown will perform various functions, such as scrolling up and down or zooming in or out.
- **Side Button** – The side button's main purpose is to bring you to the Control Center. Pressing this button will bring up the Control Center screen which allows you to quickly enable or disable some functions. Pressing and holding the side button will shut down or power up your Apple Watch.
- **Action Button** – On Ultra models only, this button can be customized to perform any action that you want.

Turning your Apple Watch On and Off

Although rarely necessary, you can power down your Apple Watch at any time. To do this, press and hold the side button until Figure 2.2 appears. Now tap the power icon at the upper right and then tap the power icon again and slide it to the right side of your screen. This will shut the Apple Watch down.

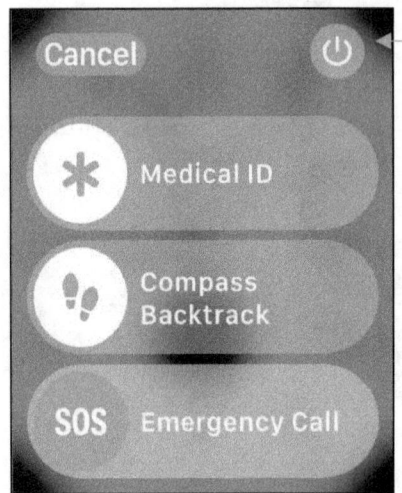

Tap to power off.

Figure 2.2 – Shutting Down the Apple Watch

To power your Apple Watch on, simply press and hold the side button until the screen lights up.

Chapter 2 | Apple Watch Layout

Waking your Apple Watch and Putting it to Sleep

Your Apple Watch will automatically dim when not in use and automatically brighten when you look at it. It will use your wrist movements to determine when you turn your watch to your face or when you lift it up.

Charging your Apple Watch

Your Apple Watch came with a magnetic charger that snaps to the back of the device. Simply plug the other end into a powered USB port or a power brick to charge your Apple Watch. Alternatively, there are magnetic stands and flat charging pads that you can purchase to charge your Apple Watch in style.

Chapter 2 | **Apple Watch Layout**

Chapter 3 | Getting Started

Chapter 3 – Getting Started

This chapter covers getting started with your Apple Watch and goes over the first-time setup procedure. If you have already completed the first-time setup where you chose your language and paired your Apple Watch with your iPhone, then you can skip the first part of this chapter.

First-time Setup – Pairing your Apple Watch

Before we begin setting up your Apple Watch for the first time, you are going to need two things: your iPhone and your Apple Watch. It is essential that you have an iPhone for your Apple Watch as it is required to use most features. Once you have both devices ready, make sure your iPhone is powered on and then we can start setting up your Apple Watch by pairing it to your iPhone.

1. Power on your Apple Watch by pressing and holding the <u>side button</u> until the display lights up.
2. When your Apple Watch powers on, it will direct you to place your iPhone directly next to it. Make sure your iPhone is powered on and place it directly next to your Apple Watch with your iPhone being to the left of the watch.
3. **On your iPhone**, a popup will appear, click <u>Continue</u>. (<u>Figure 3.1</u>)

Tap to continue

<u>Figure 3.1</u> – Pairing Apple Watch

19

Chapter 3 | Getting Started

4. **On iPhone**, tap Set Up for Myself. (Figure 3.2)

Tap to set up for yourself

Tap to set up for a family member

Figure 3.2 – Set Up Apple Watch

5. Next, your Apple Watch will show a moving graphic, and your **iPhone** will enable the camera. Take your iPhone's camera and direct it to your Apple Watch, so that the graphic on your watch appears on your iPhone's screen. You may need to move your iPhone closer to your Apple Watch to get it to connect. (Figure 3.3)

Chapter 3 | Getting Started

Align your iPhone's camera so that your Apple Watch's face comes on to your screen. Move your watch closer to camera to complete the pairing process.

Only use the pair manually option if the automatic pairing process continually fails.

Figure 3.3 – Pairing Apple Watch

6. Once your iPhone connects to your Apple Watch, it will tell you that your Apple Watch is paired.
    a. Note: If the pairing fails, I recommend you try again. If it continually fails, you can tap on Pair Apple Watch Manually and then follow the instructions on your iPhone to pair your watch.
7. Next, **on your iPhone**, select whether you want to Restore from Backup or Set Up as New Apple Watch. (Figure 3.4)
    a. If this is your first Apple Watch, tap Set Up as New Apple Watch.
    b. If this is not your first Apple Watch, you can tap Restore from Backup.

21

Chapter 3 | Getting Started

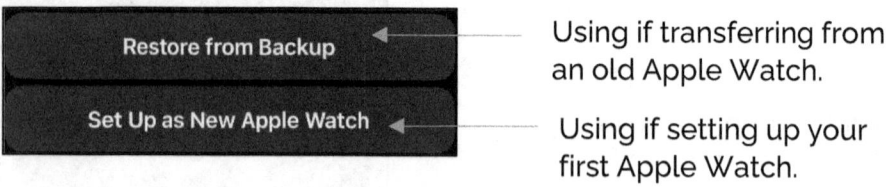

Restore from Backup — Using if transferring from an old Apple Watch.

Set Up as New Apple Watch — Using if setting up your first Apple Watch.

Figure 3.4 – Choose Setup Option

8. **On your iPhone**, you will be asked what wrist you plan to wear your watch on. Tap on the appropriate option and then tap Continue.
9. Next, **on your iPhone** read and tap Agree on the terms and conditions if you agree.
10. Next, you will need to create a passcode. A passcode is a four-digit PIN that allows you to unlock your Apple Watch. You will need to remember this PIN, so I recommend writing it down somewhere safe. You will only need to use the passcode on your Apple Watch in certain circumstances, not every time you need to access it. Once you have decided on a passcode, tap Create a Passcode **on your iPhone**. (Figure 3.5)
    a. If you prefer to create a longer passcode, you can tap add a Long Passcode **on your iPhone**.
    b. If you prefer not to use a passcode on your Apple Watch, you can tap Don't Add Passcode **on your iPhone**.

22

Chapter 3 | Getting Started

Tap to create a standard 4-digit passcode

Tap to create a long passcode

Tap to NOT create a passcode (not recommended)

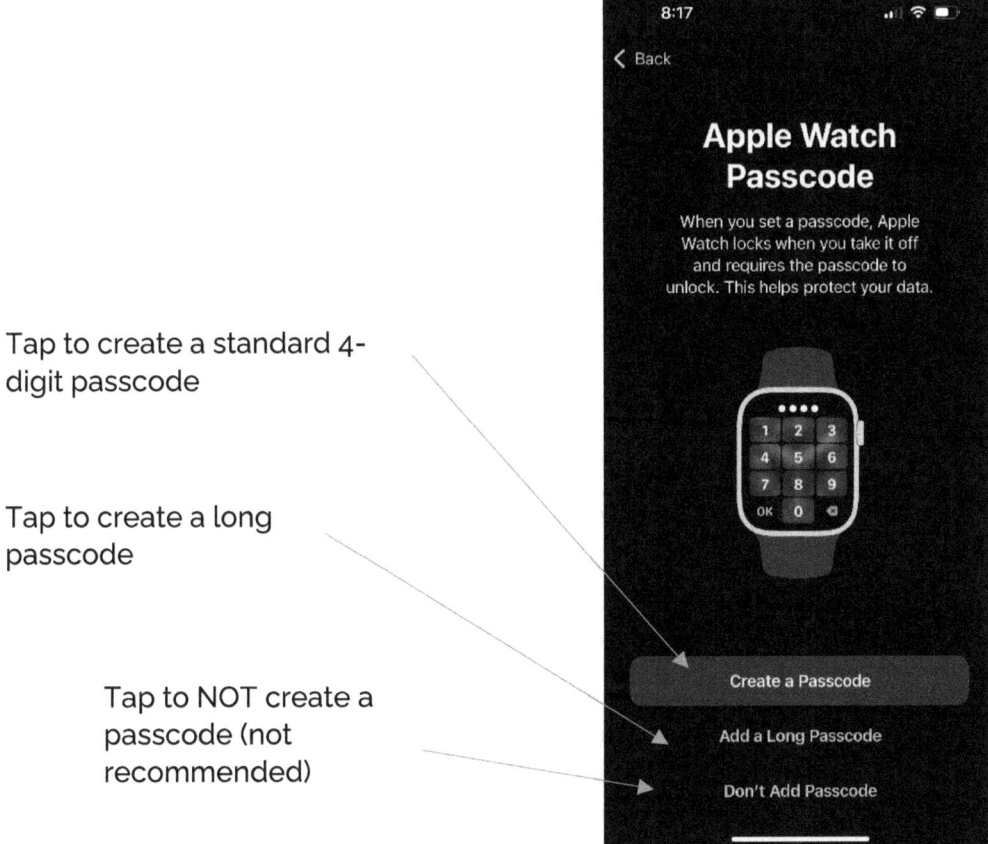

Figure 3.5 – Setting up Passcode

11. If you decide to create a passcode, **on your Apple Watch**, tap in the four-digit PIN one digit at a time. Then tap it again to confirm it.
12. After creating your passcode, **your iPhone** will ask you to select the size of the text on your Apple Watch. You can drag the slider on your iPhone to adjust, but I recommend keeping this as default for now as it can always be adjusted later. Tap Continue.
13. The next screen **on your iPhone** is the Shared Settings screen. Tap OK.
14. The next screen **on your iPhone** is the Fitness and Health screen. Tap into each box here and enter your personal information so your Apple Watch can accurately track your health and fitness metrics. When finished, tap Continue. (Figure 3.6)

Chapter 3 | Getting Started

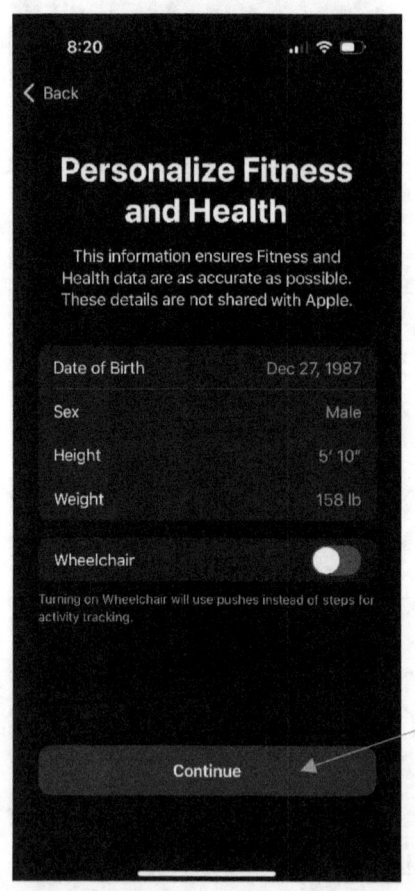

I recommend you enter your personal information here as accurately as possible. Your watch will use this information to determine your health metrics that it tracks.

Tap when finished

Figure 3.6 – Health Information

15. The next screen **on your iPhone** will be the Health Notifications screen. For now, tap Continue.
16. The next screen **on your iPhone** will be some information about safety features on your Apple Watch. Tap Continue.
17. The next screen **on your iPhone** will demonstrate how the double tap gesture works. We will cover this later so tap Continue.
18. Next, your iPhone will sign your Apple Watch into your Apple Account. This may take a few minutes.

Chapter 3 | Getting Started

19. The next screen **on your iPhone** will ask you if you would like to set up Apple Pay on your Apple Watch. Apple Pay lets you store credit card information directly on your Apple Watch, and you can then pay using these credit cards from your watch at Apple Pay enabled terminals. We will cover how to do this later so for now tap on Set Up Later.
20. Finally, the last screen **on your iPhone** will say Welcome to Apple Watch. Tap Done.
21. **On your Apple Watch**, tap Start and then tap Skip if prompted to take a tour.

After you have completed the initial setup, you will be brought to the main face on your Apple Watch, which will show you the time.

Chapter 3 | **Getting Started**

Chapter 4 | Update your watchOS Now

Chapter 4 – Update your watchOS Now

Before we get into using the Apple Watch and all its features, it is important that you check to see if an update is available to your watch's software, otherwise known as watchOS. To do this, follow these steps:

1. On your Apple Watch, press the Digital Crown to view your apps.
2. Tap the Settings app which is a gear icon. (Figure 4.1)

This vertical dashed arrow means you scan swipe up or down with your fingers or turn the Digital Crown to explore.

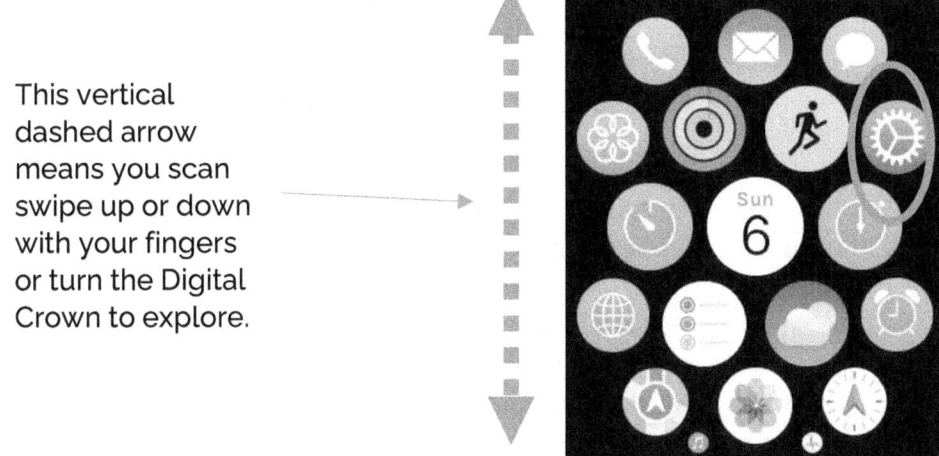

Figure 4.1 – Apps Screen -> Settings App

3. Swipe your finger up and down on the Apple Watch to scroll through the options and find General and tap it.

Chapter 4 | Update your watchOS Now

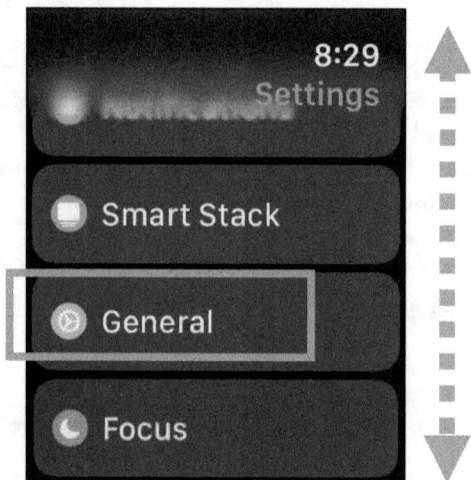

Figure 4.2 – Settings -> General

4. Tap Software Update.
5. Your Apple Watch will check to see if there is an update available. If there is an update available, your watch will say so, and there will be an option to Update Now. If this option appears, tap on Update Now. If there is not an update available, your iPhone will show "Apple Watch is up to date."
6. If an update is available, install the update by tapping on Update Now. Your Apple Watch will download and install the software update simultaneously with your iPhone. This could take some time, and you will need to be connected to Wi-Fi and have your Apple Watch connected to its charger to download the update. Let the update install, and your Apple Watch will restart when it is finished, and then you are ready to continue. If your watchOS software is already up to date, then you are ready to continue now.

You can also update your watchOS software directly from your iPhone. It is more practical to do it on your iPhone, but I want you to get comfortable using the Apple Watch's interface this early in the book. For future reference, you can update your watchOS software **from your iPhone** in the Watch App -> General -> Software Update.

Chapter 5 | Navigating your Apple Watch

Chapter 5 – Navigating your Apple Watch

Your Apple Watch is a great smartwatch that has a ton of useful features. To use your Apple Watch, all you need are your fingers. Everything is based on the touch screen and the two buttons on the side of the watch, the Digital Crown and Side Button. Here at the home face, you can see your currently selected home face which will show you various bits of information. To access any information shown on your home face, simply tap on a section by lightly touching down on the screen with your finger and quickly releasing. This will open more information about what you tapped. To go back to your home face at any time, simply press and release the Digital Crown.

The circled areas are called "complications".

You can tap on the middle of your home face to interact with it.

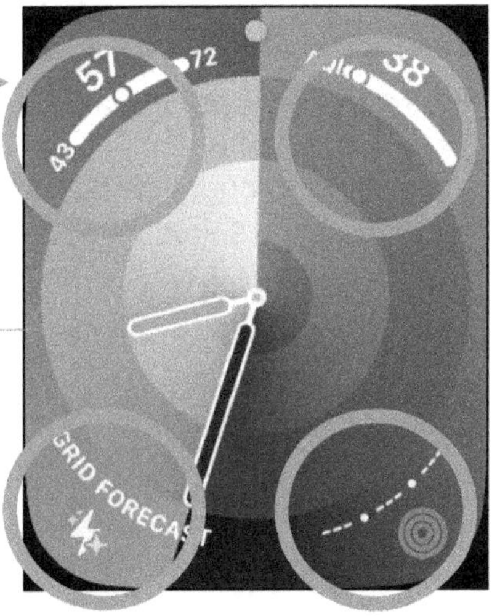

Figure 5.1 – Navigation

Digital Crown

The Digital Crown is the large button on the side of your Apple Watch that can be pressed and rotated. While on your home face, pressing the Digital Crown will bring you to your Apps Screen. While viewing the Apps Screen, you can scroll up and down by turning the Digital Crown. To return to the home face, press the Digital Crown again. (Figure 5.2)

Chapter 5 | Navigating your Apple Watch

You can scroll through your apps by turning the Digital Crown or swiping with your finger.

Simply tap on an app to open it.

Figure 5.2 – Apps Screen

When viewing your home face, rotating the Digital Crown upwards with your finger will bring up the Smart Stack, which is a set of widgets that will show you relevant information. You can tap on any widget to see more information. To leave the Smart Stack and return to the home face, simply press the Digital Crown. (Figure 5.3)

The Smart Stack shows you useful information. You can tap on any widget to open it.

Figure 5.3 – Smart Stack

The Digital Crown can sometimes be thought of as a "back button." Pressing it will bring you back a screen until you get back to your main home face.

30

Chapter 5 | Navigating your Apple Watch

Side Button

The Side Button is the rectangular button below the Digital Crown. Pressing the side button at any time will bring up the Control Center. The Control Center is where you can control many different features on your Apple Watch. We will cover the Control Center extensively later, so to return to your home face, press the Digital Crown button. (Figure 5.4)

Pressing the side button will always bring up the Control Center.

Figure 5.4 – Side Button

Faces

The faces of your Apple Watch are the main screens that show you the time and some other information. Each time you lift your Apple Watch to view it, it will illuminate and show the home face. Looking at the default face here, there is plenty of information to explore (See Figure 5.1). We can see the time, along with some information at each corner of the face. These tidbits of information on the face are called **complications**, and you can tap on any to see more information. For example, if you tap on the weather complication, it will open the Weather app on your Apple Watch (Figure 5.5). To leave the app and return to your home face, press the Digital Crown.

Chapter 5 | Navigating your Apple Watch

While viewing an app, you can browse through it by tapping on items, swiping up and down and left and right, and turning the Digital Crown.

Tapping on a complication on your home face will bring up additional information or open the associated app.

Figure 5.5 – Weather App

You can set up multiple faces on your Apple Watch and easily switch between them. In the next chapter, we will cover this.

Chapter 6 | Faces

Chapter 6 – Faces

The faces of your Apple Watch are what you will be interacting with the most, so it is important to understand this right from the beginning. In this chapter we will cover how to use and interact with the faces and how to customize them. Keep your iPhone handy as you will need it in this chapter.

The Watch App on your iPhone

Before we get into customizing your watch faces, let me introduce you to the Watch app **on your iPhone**. Open this app and your Apple Watch will appear. (Figure 6.1)

Your faces will be shown here.

All the settings of your Apple Watch can be accessed here by scrolling through the different options.

Browsing tabs are at the bottom.

Figure 6.1 – Watch App on iPhone

33

Chapter 6 | Faces

At the bottom of your screen are three tabs:

- **My Watch** – Where you can change the settings of your Apple Watch.
- **Face Gallery** – Where you can browse for and add faces to your Apple Watch.
- **Discover** – Where you can read about certain features of the Apple Watch.

We will be using the Watch app extensively in this book, as it the most practical way to change various settings on your Apple Watch.

Adding Faces to Apple Watch

Stay in the Watch app **on your iPhone** and let us see how we can add faces to your Apple watch and then customize them. To get started, tap the Face Gallery tab at the bottom of the Watch app. (Figure 6.2)

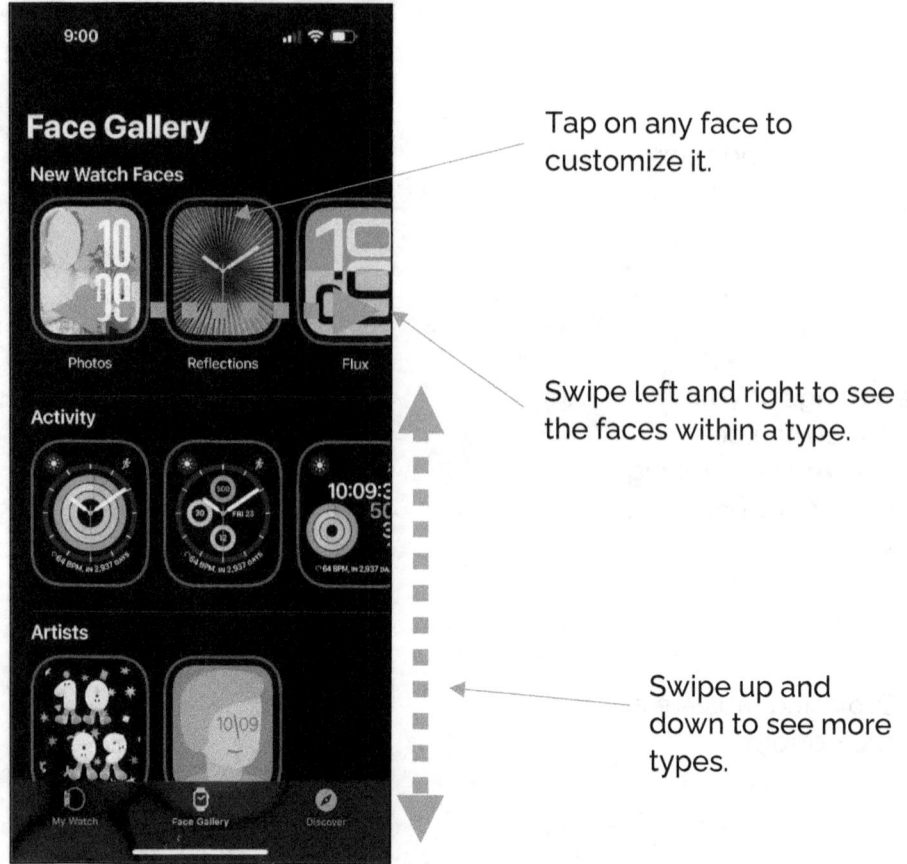

Tap on any face to customize it.

Swipe left and right to see the faces within a type.

Swipe up and down to see more types.

Figure 6.2 – Face Gallery on iPhone

Chapter 6 | Faces

Face Gallery

Here inside the Face Gallery, you can browse through all the different faces available. You can swipe up and down and left and right inside the Face Gallery to see all the different types of faces. Faces are organized by type, and I will explain the different types to you:

- **New Watch Faces** – Shows the newest Apple Watch faces available.
- **Activity** – These faces focus on the health and exercise features of your Apple Watch. They often will show you how many steps you have taken today, your move goals, and offer quick access to fitness apps.
- **Artists** – These faces are art-based with no particular focus.
- **Astronomy** – The astronomy faces focus on the moon, earth, and solar system. With these faces, there is often a stunning graphic of a planet or solar system customized to the exact time of day.
- **Breathe** – The Breathe faces are focused on mindfulness. You can set up breathing exercises on your Apple Watch and the Breathe faces will quickly bring you to them. In addition, these faces will often show you mental health and heart rate information.
- **California** – The California faces are styled after California-style watches. There are light and dark themed faces and there is no particular focus for these. These faces can be highly customized.
- **Chronograph Pro** – These faces are styled after chronograph watches. They frequently show time-based information such as quick access to a stopwatch, timer, and a tachymeter.
- **Color** – Color faces focus on the color schemes. These are highly customizable so you can create a face that emphasizes your choice of colors.
- **Contour** – The contour face is a special face that gradually changes to highlight the current hour.
- **Count Up** – These faces are focused on elapsed time. They are great for setting up custom timers on the go.
- **Fire and Water** – These faces are animated graphics which depict fire and water.
- **Flux** – Flux faces are styled based on font and color, with both frequently changing based on the minute.

Chapter 6 | Faces

- **GMT** – The GMT faces are focused on local vs Greenwich Mean Time. You can customize these to show two different time zones at the same time.
- **Gradient** – These faces allow you to set gradients as the color that changes with time.
- **Infograph** – Infograph faces are all about information. These can be highly customized to show up to 8 different complications.
- **Kaleidoscope** – These faces are based on the imagery of a kaleidoscope. You can see a kaleidoscope active effect while using these faces by turning the Digital Crown.
- **Liquid Metal** – These faces are graphically based on the materials that are used to make the Apple Watch.
- **Lunar** – This face is dedicated to the moon. It can be used to depict the date to the phase of the moon.
- **Memoji** – These animated faces feature any Memojis you have created. Kids love to play with this face.
- **Meridian** – Similar to the Infograph face, but with less complications. A standard style face.
- **Metropolitan** – A classic style face with a type-driven watch.
- **Mickey and Minnie Mouse** – These animated faces depict Mickey and Minnie Mouse. You can tap on the face when active and Mickey or Minnie will speak the time to you. Popular with children.
- **Modular, Modular Compact, and Modular Duo** – Digital style faces with up to 6 complications that are highly customizable.
- **Motion** – These animated faces show motion when viewed and can show jellyfish, butterflies, or flowers. Every time you raise your wrist a different iteration will be shown.
- **Nike Faces** – These faces are based off the Nike Run Club app.
- **Numerals Mono & Duo** – The Numerals faces display large numbers for easy viewing.
- **Palette** – Palette uses dynamic colors to show different aspects of the watch face. The colors will change as the second hand makes its way around the watch.
- **Photos** – These popular faces will pull photos from the Photos app on your iPhone and display them as a face on your Apple Watch. You can choose

Chapter 6 | Faces

specific photos or albums that you want to show or let the Apple Watch choose for itself.
- **Pride** – Pride based faces.
- **Reflections** – These faces offer a graphical reflection of light based on the movement of your wrist.
- **Simple** – The simple faces are highly customizable to show you exactly what you want to see.
- **Snoopy** – This face has an animated Snoopy character that can be interacted with. Popular with children.
- **Solar & Solar Analog** – The solar faces track the sun's position throughout the day and relate it to the current time.
- **Stripes** – These faces can be highly customized to show different color stripes across the screen.
- **Timelapse** – These faces show a location that updates based on the current time. For instance, the New York location will show daytime New York during daylight hours and nighttime New York in the evening.
- **Toy Story** – Toy Story based face that is popular with children.
- **Typograph** – Font-based faces that focus on typography.
- **Unity** – Black unity faces.
- **Utility** – These faces are focused on practicality and can be highly customized.
- **Vapor** – Graphically animated faces focused on colors.
- **World Time** – Classical style faces that focus on times around the world.
- **X-Large** – These simple faces display the time in extra-large font for easy viewing.

This is not a complete list of all the face categories as they are frequently updated.

Adding a Face to your Apple Watch

Adding a face to your watch from the Face Gallery is simple. **On your iPhone**, follow these steps:

1. Open the Watch app.
2. Tap Face Gallery.
3. Find a face you want to add to your Apple Watch and tap *it*.

Chapter 6 | **Faces**

4. On the next screen, you can customize the face before adding it to your watch. (Figure 6.3)
    a. The color options allow you to set the color scheme of the face.
    b. The Complications section allows you to change the complications on the face. Remember, complications are small widgets that appear on the face, such as weather, activity, and other apps. You can tap on each customization to change it.
5. To add the face to your Apple Watch, tap Add.

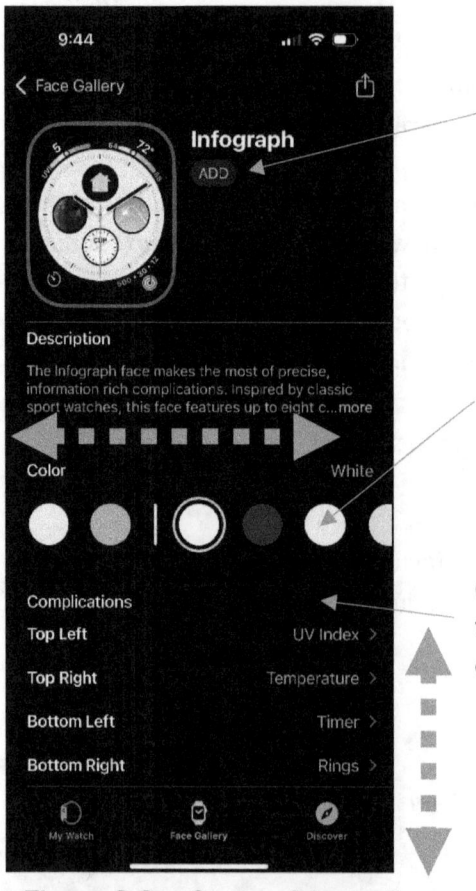

Add the face to your Apple Watch and set as home face.

Set the color scheme of the watch. You can swipe left and right to see more.

Change the complications of the face, which are the widgets on each corner of the face.

Figure 6.3 – Customizing a Face on iPhone

6. The face will now be added to your Apple Watch and will be automatically set to display it.

38

Chapter 6 | Faces

In the above example, we added the Infograph face to our Apple Watch (Figure 6.4). We did not edit any of the complications. Let's look at another example and this time, we will do some more customization.

Figure 6.4 – New Apple Watch Face

Example: Adding a Face
1. Open the Watch app **on your iPhone**.
2. Tap Face Gallery.
3. Scroll down until you find Meridian and tap *on any face here* to customize it. (Figure 6.5)

Chapter 6 | Faces

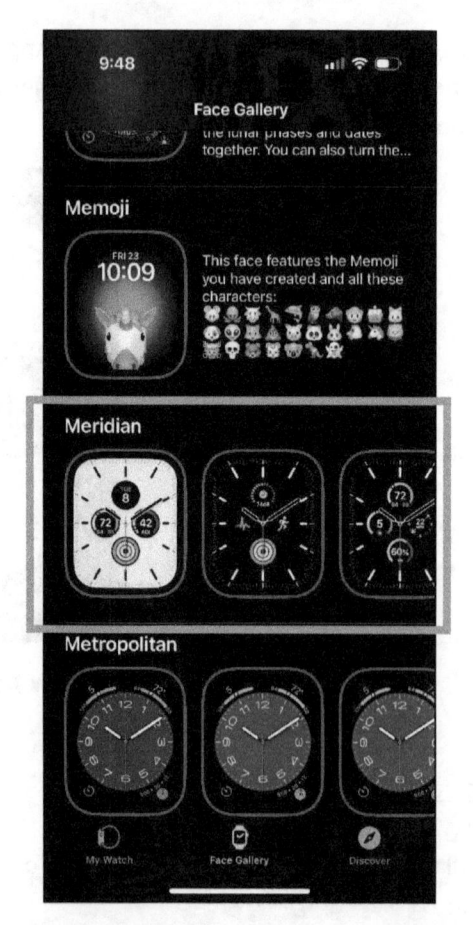

Figure 6.5 – Face Gallery -> Meridian

4. For this face, we can customize both the color of the complications and the color of the dial. (Figure 6.6)

Chapter 6 | Faces

Any changes to the face will be shown here.

Change the color scheme.

Change the dial color.

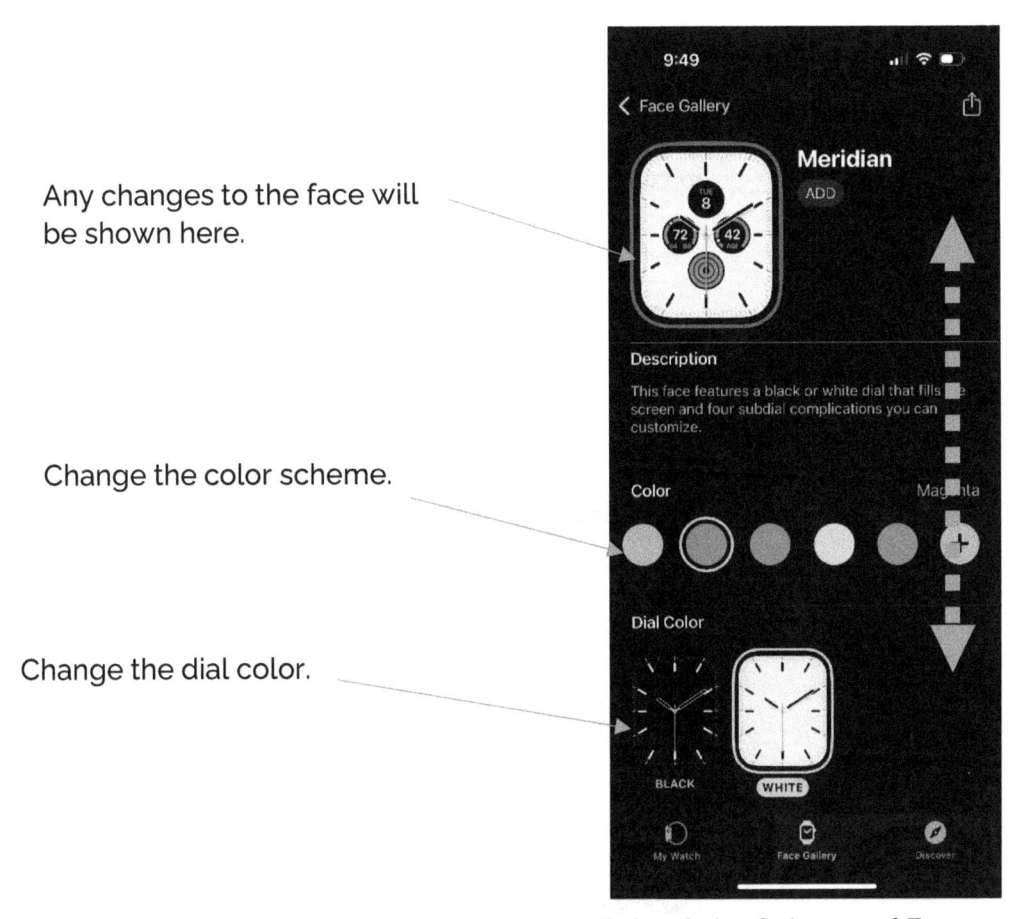

Figure 6.6 – Color Scheme of Face

5. Scrolling down, we can change each complication on the face.
6. You can tap on each location and then browse through all the complications available to change them.
7. When finished, tap ADD.
8. The face will be added to your Apple Watch and automatically set to it.

Complications

Complications are small widgets that are on your watch faces. These complications display information to you, and if you tap on them, will show you more information or bring you to an app. In this section, I will show you how complications work and how you can customize faces with them.

41

Chapter 6 | **Faces**

To explore the different complications available, we need to either edit a face or add a new one to the Apple Watch. To make things simple, let us add a new face to the Apple Watch **on your iPhone:**

1. Open the Watch app **on your iPhone**.
2. Tap Face Gallery.
3. Find the Modular category and tap *any face here* to begun customizing it. (Figure 6.7)

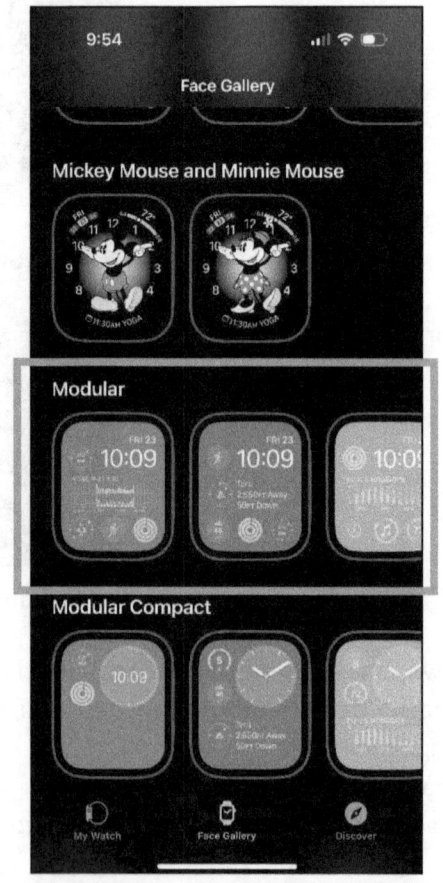

Figure 6.7 – Face Gallery -> Modular

4. As usual, you can customize the color scheme and background with the options near the middle.

Chapter 6 | Faces

5. Scrolling down, you can also customize the appearance of the numerals on this face.
6. Keep scrolling down until you get to the Complications section. (Figure 6.8)

Any changes to the face will be shown here.

All the different complications can be customized here.

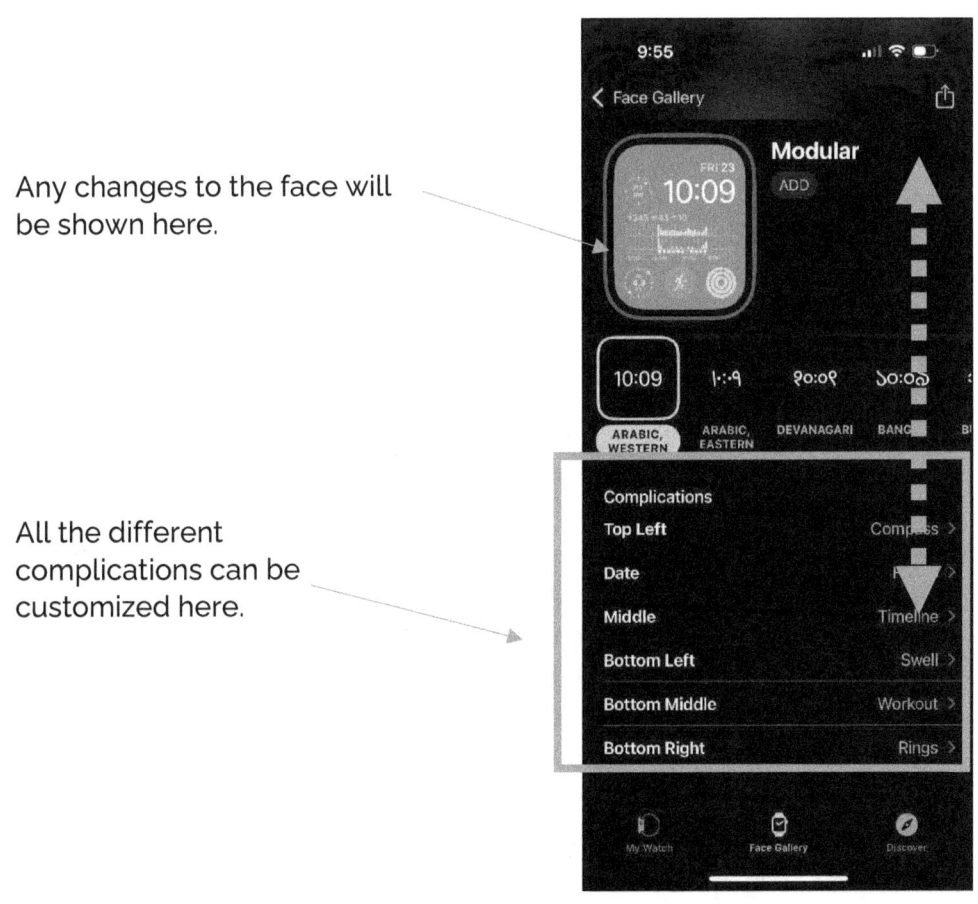

Figure 6.8 – Complications

7. There are many different complications for this face, and each one can be changed to the option of your choosing. To change any of the complications on the face, tap on its *location*.
8. The next screen will show you all the complications available. The list of available complications will depend on what apps you have installed on your iPhone. By default, there are many useful complications that come preloaded:

Chapter 6 | Faces

- **Activity** – This complication is a graphic that shows you the status of your activity goals.
- **Alarms** – This complication will open your alarm clock.
- **Astronomy** – The various choices here allow you to set a graphic, such as the Earth, Moon, and Sun.
- **Battery** – The battery complication allows you to display your battery life.
- **Blood Oxygen** – This complication allows you to see your most recent Blood Oxygen reading or to take a reading right now.
- **Calculator** – Opens the Calculator app.
- **Calendar** – Either shows you the current date or your schedule for the day.
- **Camera** – Opens the camera remote, which allows you to capture a photo on your iPhone from your Apple Watch.
- **Compass** – Displays the compass.
- **Contacts** – Allows you to display a specific contact that can easily be brought up.
- **Cycle Tracking** – Displays your cycle tracking information for women's health.
- **ECG** – Allows you to open the ECG app, which takes an ECG.
- **Heart Rate** – Allows you to take your heart rate or will display your most recent heart rate measurement.
- **Messages** – Allows you to quickly open the Messages app.
- **Music** – Allows you to open the Music app.
- **Phone** – Allows you to open the Phone app.
- **Reminders** – Allows you to open the Reminders app to see your lists.
- **Sleep** – Allows you to see your most recent sleep data.
- **Stocks** – Displays stock information.
- **Tide** – Displays current tide information.
- **Time** – Displays different time information.
- **Timer** – Allows you to quickly start a timer.
- **Wallet** – Allows you to access your Apple Pay wallet.
- **Weather** – Allows you to display specific weather information.
- **World Clock** – Allows you to display different time zones.

You can tap on any of these complications to add it to the face.

Chapter 6 | Faces

9. Once you have finished changing the complications, tap ADD and the face will be added to your Apple Watch.

Editing or Deleting a Face that has been Added to your Apple Watch

You can easily edit a face that you have already added to your Apple Watch by using the Watch app **on your iPhone**.

1. Open the Watch app **on your iPhone**.
2. Tap the My Watch tab.
3. At the top, all your faces will be shown. Top on one to edit it. (See Figure 6.1)
4. While editing, you can change all the aspects of this face, including the color scheme and complications. Any changes you make will automatically apply on your Apple Watch.
5. To delete a face from your watch, scroll down and tap Remove Watch Face.

Switching Between Faces on Apple Watches

Switching between your faces on your Apple Watch is very simple. To do this:

1. Tap and hold on to your current face until your watch vibrates and your screen separates. (Figure 6.9)

Swipe left and right to switch between faces.

Tap on a face to set it as your home face.

Tap to edit a face directly on your Apple Watch.

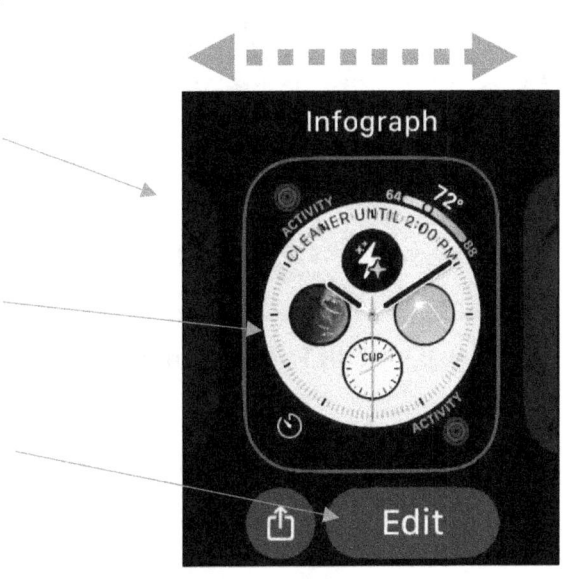

Figure 6.9 – Switching Faces

45

Chapter 6 | Faces

2. Swipe to the left or right to select another watch face.
3. Tap on the _face_ to select it and set it as your home face.

Adding or Editing Faces Directly on the Apple Watch

Instead of using the Watch app on the iPhone to add or edit faces, you can do it directly on your Apple Watch. This method is not recommended as it is generally more tedious to do, but I will show you how regardless.

1. **On your Apple Watch**, tap and hold on _your current face_ until your watch vibrates and the screen separates. (See Figure 6.9)
2. To edit the current face, tape Edit. (See Figure 6.9)
    a. Use the Digital Crown and the screen to edit the different aspects of the face. (Figure 6.10)

Although it is easier to edit a watch face on your iPhone in the Watch app, you can edit a face directly on your Apple Watch.

Use the Digital Crown and your fingers to customize the face.

Figure 6.10 – Edit Face on Apple Watch

   b. You can swipe left and right to switch between editing the color scheme and complications. (Figure 6.11)

46

Chapter 6 | Faces

Tap on a complication and then turn the Digital Crown to change it.

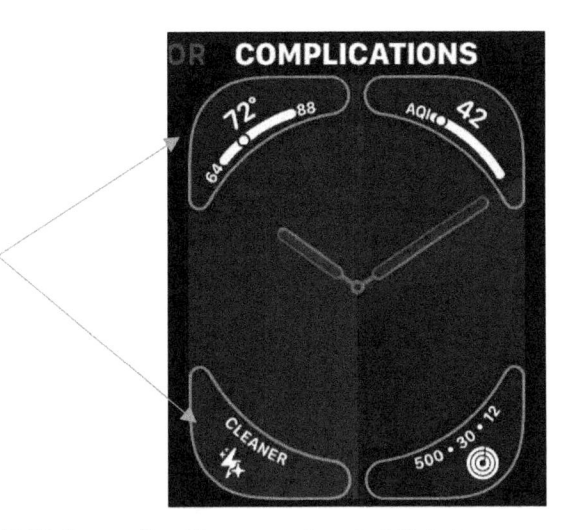

Figure 6.11 – **Edit Complications on Apple Watch**

3. To add a new face, swipe your finger to the left repeatedly until you find the plus symbol and tap it.
    a. Now you can browse the Face Gallery directly on your Apple Watch to add one.

That pretty much covers all you can do with faces on your Apple Watch. We only sampled a few of the many faces available. It is worth experimenting with the many different faces available in the Face Gallery and seeing exactly what they look like on your Apple Watch. As we continue through this book, it will become clearer which apps you would like to use and what complications you will prefer on your faces.

47

Chapter 6 | **Faces**

Chapter 7 | The Control Center

Chapter 7 – The Control Center

The Control Center is an integral feature on the Apple Watch that allows you to control some essential functions.

Accessing the Control Center

To access the Control Center on your Apple Watch, simply press the side button at any time. (Figures 7.1 through 7.3)

Here in the Control Center, there are many buttons you can tap on to enable or disable certain features. You can access additional features by swiping your finger up and down or by turning the Digital Crown. When a function is highlighted, that means it is currently enabled.

Wi-Fi

Battery

Flashlight

Airplane Mode

Find iPhone

Focus Mode

Figure 7.1 – Control Center 1

49

Chapter 7 | The Control Center

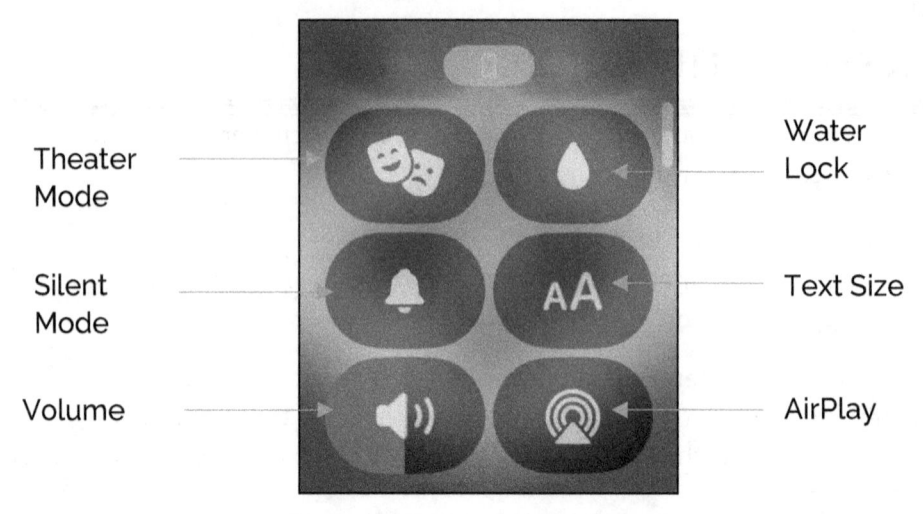

Theater Mode — Water Lock

Silent Mode — Text Size

Volume — AirPlay

Figure 7.2 – Control Center 2

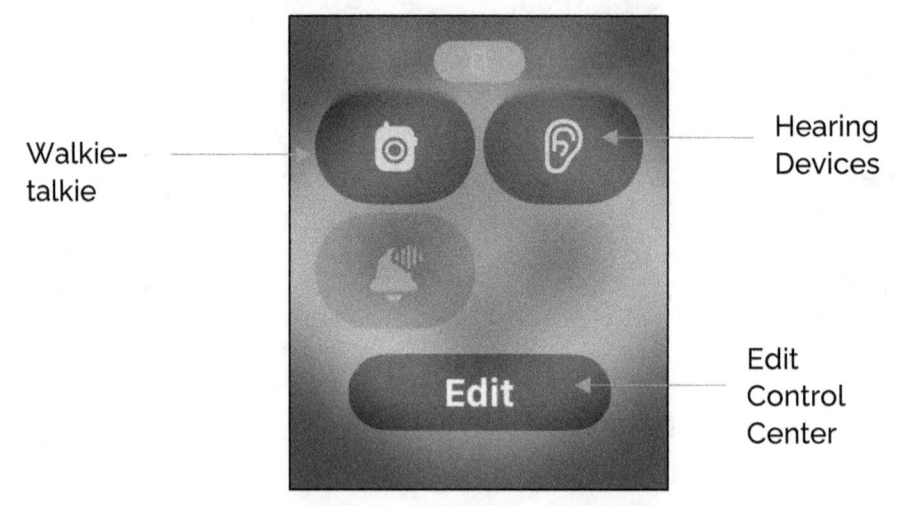

Walkie-talkie — Hearing Devices

Edit Control Center

Figure 7.3 – Control Center 3

Control Center Functions

- **Cellular** (Cellular Models Only) – Enables or disables cellular connection
- **Wi-Fi** – Tapping on this icon enables or disables Wi-Fi on your Apple Watch. Your Apple Watch connects to your iPhone via Bluetooth and shares your iPhone's Wi-Fi connection.

Chapter 7 | The Control Center

- **Airplane Mode** – Tapping on this icon enables or disables Airplane Mode. When Airplane Mode is enabled, all communication abilities of the Apple Watch become disabled, including Wi-Fi and cellular connections.
- **Battery Percentage** – This icon displays the current battery life of your Apple Watch. Tapping on it brings up more information including the option to enter Low Power Mode.
- **Find My iPhone** – This icon pings your iPhone to help you find it. Once you tap this icon, your iPhone will play a loud sound. In addition, your Apple Watch will open a tracking feature to help hone in on your iPhone's location. This is a great feature to help you locate your iPhone if you misplaced it somewhere nearby.
- **Flashlight** – This icon turns on the flashlight on your Apple Watch. The flashlight feature turns the face of your Apple Watch completely white and increases its brightness, allowing you to see in dark places. You can adjust the brightness by turning the Digital Crown and switch between colors by swiping left and right. Press the Digital Crown to exit the flashlight.
- **Moon Icon** – This icon allows you to enter a focus mode, such as Do Not Disturb, Work, and Sleep. When a focus mode is enabled, it changes the notification and sound settings of both your Apple Watch AND your iPhone. When Do Not Disturb is enabled, both your Apple Watch and iPhone will not make any sounds for notifications when you are not actively using it.
- **Theater Mode** – Theater Mode is a special mode specifically designed for the Apple Watch. When enabled, it will keep your screen dark and silence all sounds. To see your screen, you will need to turn the Digital Crown.
- **Water Lock** – Enabling water lock disables the touch screen on your Apple Watch and provides some protection for your Apple Watch when using it under water. You should enable Water Lock if you are going swimming with your Apple Watch on. In fact, the Apple Watch is made to handle underwater use. To turn off Water lock, you must press and hold the <u>Digital Crown</u> until your watch notifies you that it has been unlocked. During this time, your watch will emit a tone to eject water that might have accumulated on the speakers.
- **Silent Mode** – Enabling or disabling silent mode turns sounds on or off for your Apple Watch.

Chapter 7 | The Control Center

- **Text Size** – Allows you to quickly change the size of text on your Apple Watch. (Figure 7.4)

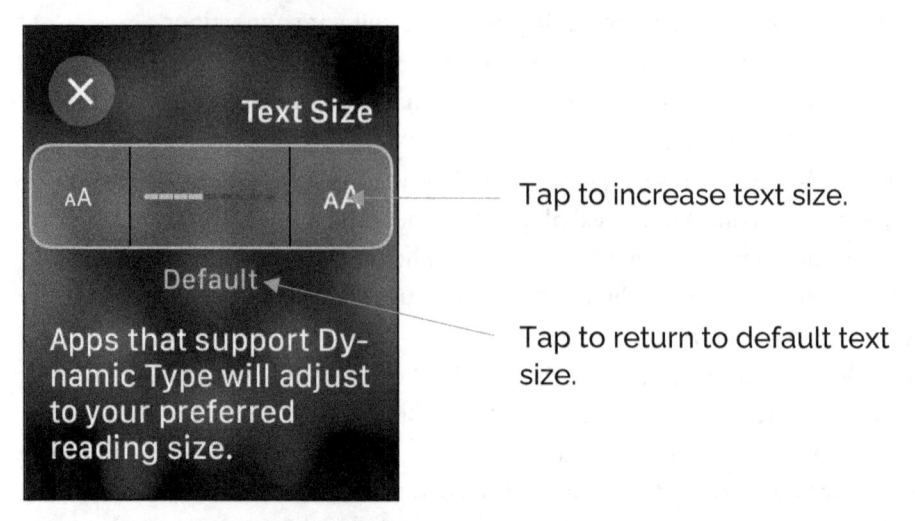

Tap to increase text size.

Tap to return to default text size.

Figure 7.4 – Adjusting Text Size

- **Volume** – Allows you to adjust the volume on your Apple Watch. Slide your finger up or down to adjust this. (Figure 7.5)

Tap and slide your finger up or down to adjust volume.

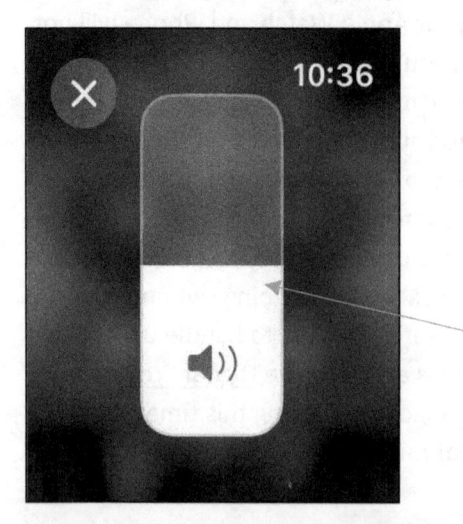

Figure 7.5 – Adjusting Volume

52

Chapter 7 | The Control Center

- **AirPlay** – Allows you to set an AirPlay device to your Apple Watch. AirPlay allows you to share your Apple Watch screen with a compatible device, such as an Apple TV.
- **Walkie-Talkie** – Enables or disabled Walkie-Talkie, which allows you to exchange live voice messages with other Apple Watch users.
- **Hearing Devices** – Allows you to access and change some settings on hearing devices you have paired with your Apple Watch, such as Apple AirPods.

You can also add or remove functions from the Control Center by tapping Edit at the bottom. To leave the Control Center, simply press the side button again.

53

Chapter 7 | **The Control Center**

Chapter 8 – Notifications

Notifications are an integral part of the Apple Watch and are an aspect you will frequently be using. To define, a notification is simply specific information that is delivered to you from an app. Any app can send you a notification, and they can pertain to just about anything. Examples of notifications include new text messages, a phone call, a breaking news alert, an activity alert, or a new email.

Your Apple Watch works seamlessly with your iPhone to deliver notifications to you in the best possible manner. For instance, while wearing your Apple Watch and not looking at your iPhone, a new notification will be sent directly to your Apple Watch, and your watch will gently tap your wrist and play a chime to notify you about it. In this case, your iPhone will not play a chime and will not vibrate since your Apple Watch decided to give you the notification.

When your Apple Watch taps you to tell you have a notification, you can lift your watch and look at it to view the notification. You can tap on the notification to go directly to the app the notification is from. For instance, if you receive a notification of a new text message, tapping on it will open the Messages app on your watch and allow you to reply. If you do not tap on the notification, after a few moments the notification will disappear. Alternatively, you can swipe a notification upwards to dismiss it.

The Notification Center
You can access all your recent notifications on your Apple Watch in the Notification Center. To access the Notification Center, tap at the top of your screen and swipe down. (Figure 8.1)

Here all your notifications will be listed. You can tap on any to see more information or go to its specific app. You can clear a single notification by swiping to the right and then tapping the x. You can clear all your notifications by swiping to the top of the list and then tapping Clear All. (Figure 8.2)

To leave the Notification Center, simply press the Digital Crown.

Chapter 8 | Notifications

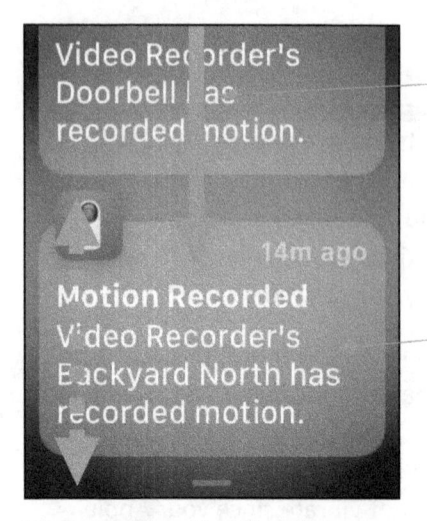

Tap at the top of your screen and swipe down to bring up the Notification Center.

Tap on a notification to see more information.

<u>Figure 8.1</u> – Notification Center

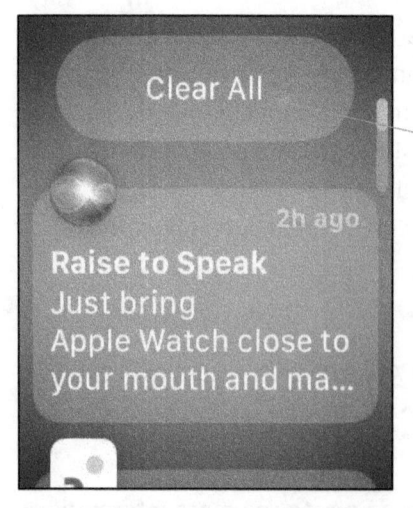

Scroll all the way up and tap <u>Clear All</u> in the Notification Center to dismiss all notifications.

<u>Figure 8.2</u> – Clear All Notifications

Some notifications will not be able to deliver you complete information on your Apple Watch, such as notifications from some third-party apps. For these notifications, you can view more information about them on your iPhone.

With the conclusion of Chapter 8, we have covered all the core features of the Apple Watch. The next few chapters will cover the various apps on the Apple Watch and what features they bring to your device.

Chapter 9 | Text Messaging

Chapter 9 – Text Messaging

You can view and send text messages directly from your Apple Watch. Your watch is particularly excellent at viewing text messages, especially through notifications which we covered in the last chapter. When you receive a new text message, your watch will notify you and you can raise your watch towards your face to quickly read it. If you want to, you can reply straight from your Apple Watch.

Sending Text Messages and the Messages App

It is a far easier experience to **send** text messages from your iPhone. However, there are some circumstances where sending text messages from your Apple Watch can come in handy. Personally, I have used my Apple Watch to send text messages when I am in a situation where I do not want to be seen on my phone. I have also used my Apple Watch to send texts when my iPhone is not nearby or if I only need to send a quick response. Regardless, you can do all of this from the Messages app.

To access the Messages app:

1. While viewing your face on the Apple Watch, press the Digital Crown to open the Apps Screen.
2. Find and tap the Messages app to open it. (Figure 9.1)

Tap to start a new text message thread.

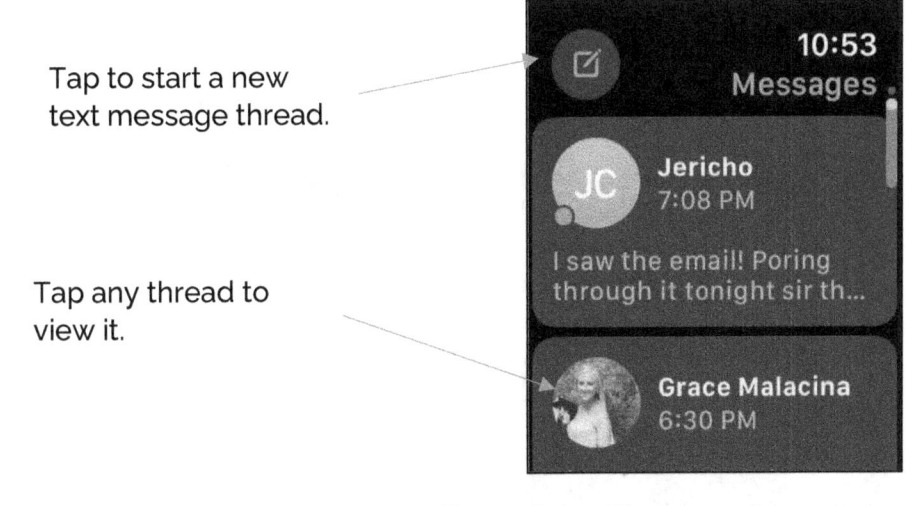

Tap any thread to view it.

Figure 9.1 – Messages App

Chapter 9 | Text Messages

Here you can view all your text message threads, including pinned threads. To open any thread, simply tap on *it*.

While inside a thread, you can read the contents by swiping up or turning the Digital Crown (Figure 9.2).

Swipe down to see the thread's history.

Tap to enter a message manually or via dictation.

Swipe up to see quick responses.

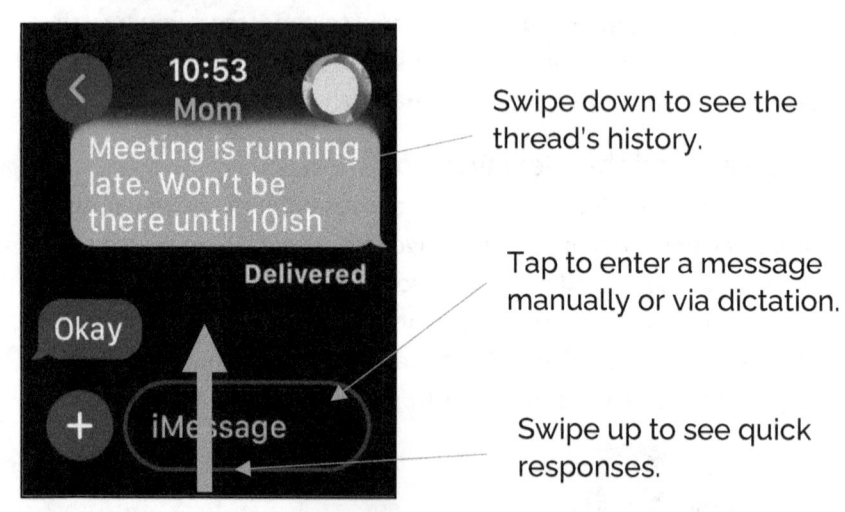

Figure 9.2 – Reading a Thread

To send a message, you can either tap into the message bar or send a quick response. To send a quick response, scroll down until you see the recommended responses and you can tap on any one to immediately send it. (Figure 9.3)

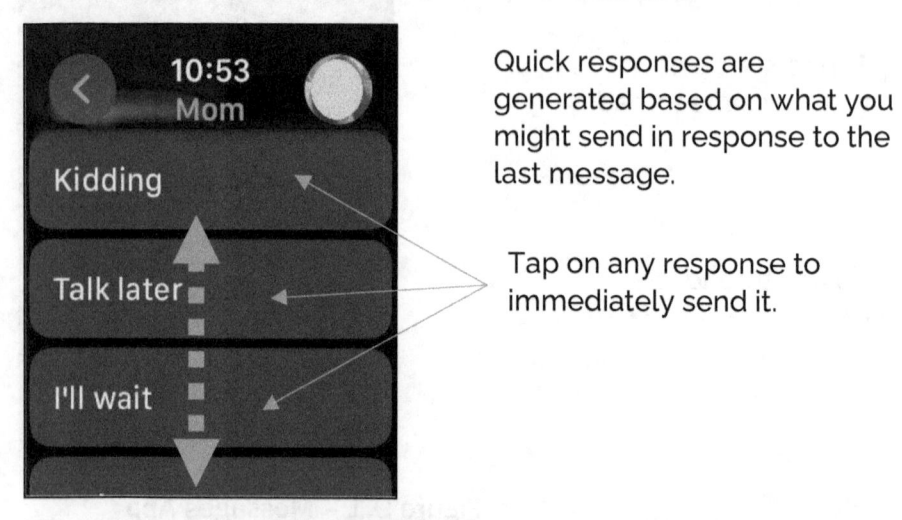

Quick responses are generated based on what you might send in response to the last message.

Tap on any response to immediately send it.

Figure 9.3 – Quick Responses

Chapter 9 | Text Messaging

TIP: When viewing a notification of a text message, you can tap on it to see quick responses that you can send. Your Apple Watch will automatically generate quick responses that it thinks you might want to use.

If none of the quick responses are what you are looking for, you can send a detailed message by tapping into the *message bar*. (Figure 9.4)

Dictate your message.

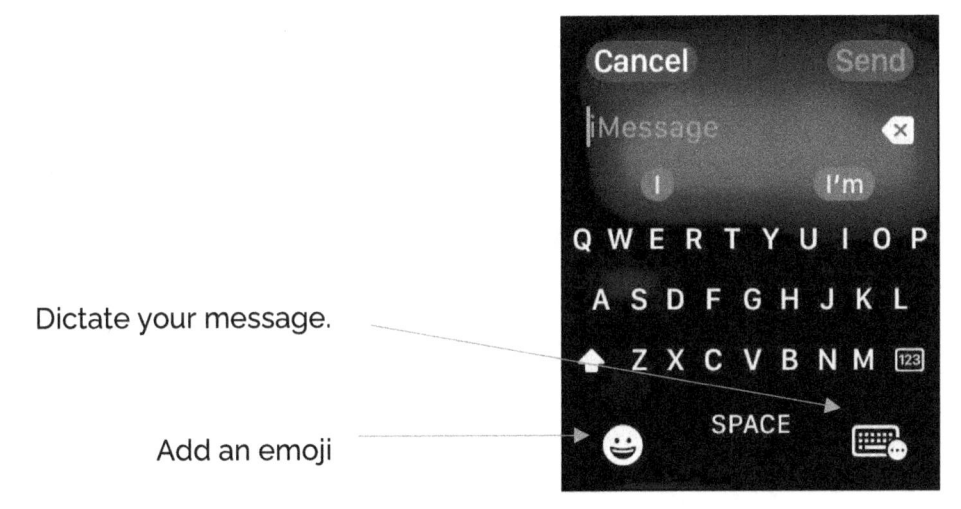

Add an emoji

Figure 9.4 – Sending Text

On this screen, you can have a couple of different options. You can use the keyboard to enter the text message directly, or you can dictate to your Apple Watch exactly what you want to send. To use the keyboard, simple tap in your message. To dictate, tap the keyboard icon and then tap the microphone icon. Now you can dictate your message to your Apple Watch and then tap Send when finished.

59

Chapter 9 | **Text Messages**

Chapter 10 | Phone Calls

Chapter 10 – Phone Calls

You can use your Apple Watch to make and receive phone calls. Doing so is quite simple.

Receiving a Call
When you receive a call, your Apple Watch will notify you with a notification. If you'd like, you can answer the call directly from your watch by tapping on the green phone icon. (Figure 10.1)

Answer the call on Apple Watch.

More options

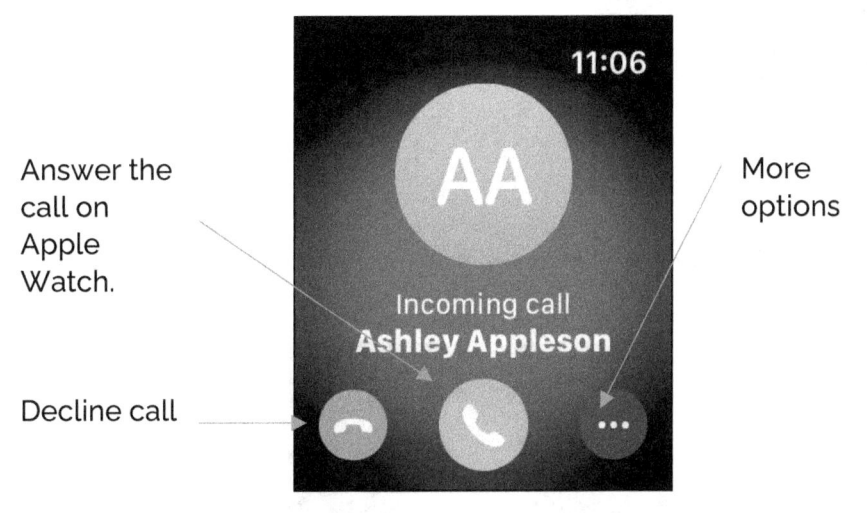

Incoming call
Ashley Appleson

Decline call

Figure 10.1 – Receiving a Call

When receiving a call on your Apple Watch, the call will play through your watch's speakers, and you can speak directly through your watch's microphone.

Making a Call
You can also initiate a call from your Apple Watch. To do this, open the Phone app from the Apps Screen.

From here, browse for the contact you wish to call. To enter a number directly, tap the keypad icon. (Figure 10.2)

61

Chapter 10 | **Phone Calls**

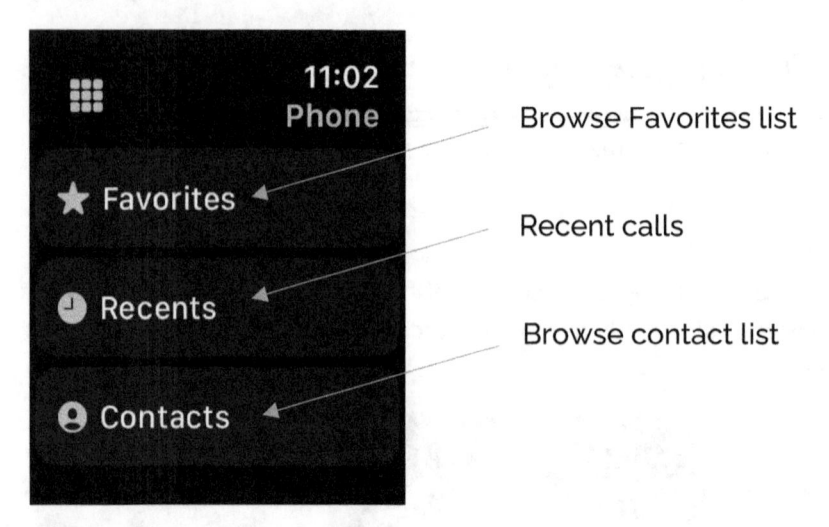

Figure 10.2 callouts:
- Browse Favorites list
- Recent calls
- Browse contact list

Figure 10.2 – Phone App

Call Functions

While on a call with your Apple Watch, there are many functions available to you (Figure 10.3):

Figure 10.3 callouts:
- End the call
- Mute
- More options
- 00:01 Ashley Appleson

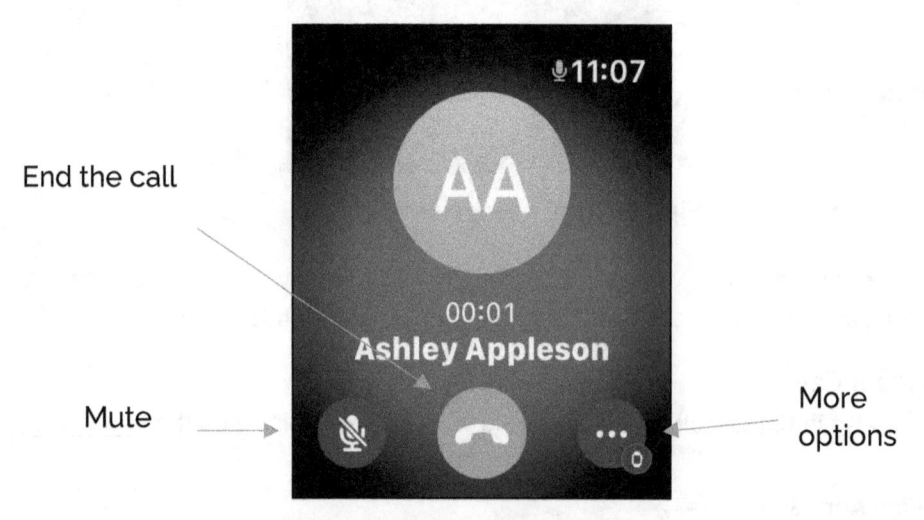

Figure 10.3 – Call Options

- **Mute** – Mutes your side of the call, i.e., the other person will not be able to hear you until you turn mute off.
- **End Call** – Ends the call

62

Chapter 10 | Phone Calls

- **More Options**
    - **Keypad** – Access the keypad for touch-tone features.
    - **Audio** – Allows you to transfer the call to another device, such as your iPhone or AirPods.

Tips for Using Phone Calls on your Apple Watch

It is usually more practical to answer a call on your iPhone rather than on your Apple Watch, however there are many circumstances where I have found the ability to answer phone calls from my watch to be very useful.

- While exercising, especially when wearing AirPods. The Apple Watch will automatically send the call to your AirPods.
- When your iPhone is not nearby.
- When your hands are not free and grabbing your iPhone is just not practical.

Chapter 10 | **Phone Calls**

Chapter 11 | Activity App

Chapter 11 – Activity App

The Activity app is a great health feature built into the Apple Watch which tracks your movement and activity. To get started with this, open the Activity app from the Apps screen.

Using the Activity App
Your activity goals will automatically be set up based on your age, sex, height, and weight that you first provided to your Apple Watch during the initial setup. The Activity app will track your progress towards these goals. (Figure 11.1)

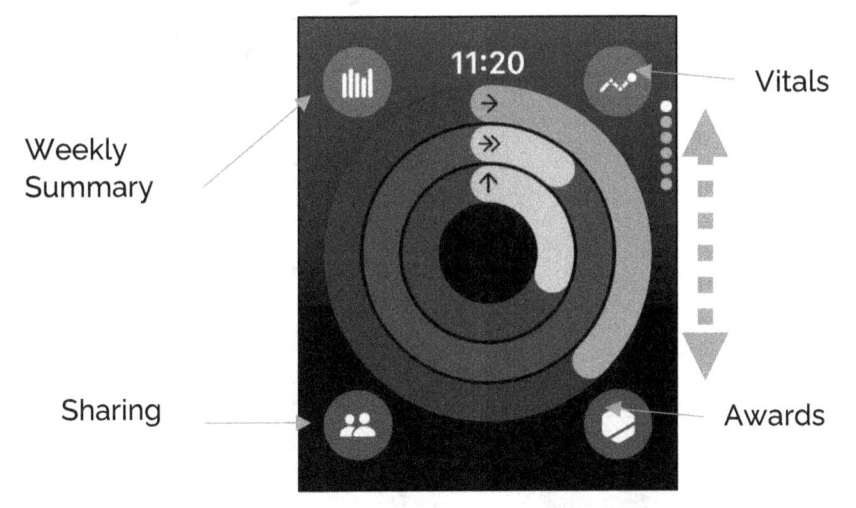

Figure 11.1 – Activity App

The Activity app uses three rings to show your progress. The red ring is your move goal and tracks the calories you have burned simply by moving throughout the day (Figure 11.2). The green ring is your exercise goal, and tracks how many minutes you have exercised for the day (Figure 11.3). Lastly, the blue ring is your stand goal, and tracks how many times you have stood up during each hour of the day (Figure 11.4)

Chapter 11 | Activity App

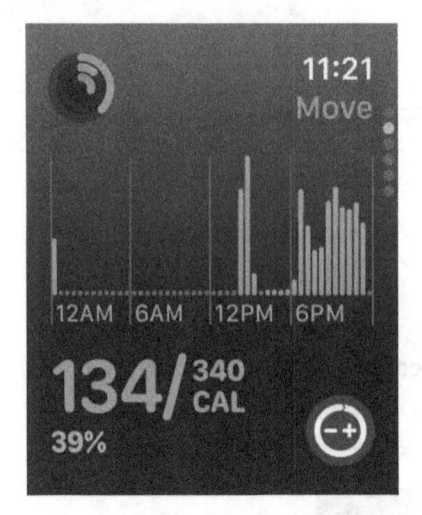

Figure 11.2 – Move Goal

Figure 11.3 – Stand Goal

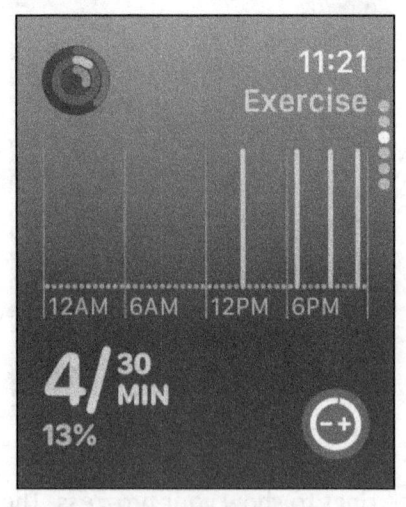

Figure 11.4 – Exercise Goal

By continuing to scroll down in the Activity app, you can view more information including the steps you have taken today, the total distance you have moved, and the number of flights you have climbed.

All of this is tracked automatically using both your Apple Watch and iPhone in sync with each other. The Activity app also integrates with the health features of your Apple Watch, which we will get into soon.

Chapter 12 | Fitness & Exercise

Chapter 12 – Fitness & Exercise

Your Apple Watch is an excellent tool for monitoring and tracking your exercise routines. The watch has built in sensors to measure key metrics for several different exercises. All these exercises can be found in the Workout app in the Apps Screen.

Workout App

Here in the Workout app, you can set your Apple Watch to track your exercises. (Figure 12.1) Your watch can track all sorts of exercises, including: outdoor walk, outdoor run, indoor run, outdoor cycle, indoor cycle, pool swim, open water swim, multisport, hiking, elliptical, stair stepper, indoor rowing, high intensity interval training, kickboxing, functional strength training, core training, yoga, Pilates, dance, Tai Chi, cooldown, and more.

Tap any workout to start it.

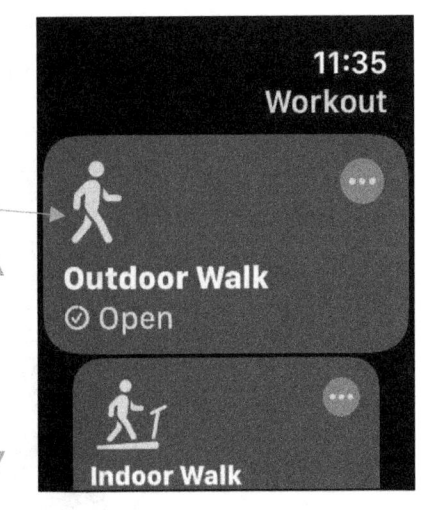

Figure 12.1 – Workout App

To begin tracking a workout, in the Workout app simply tap on _the workout you are about to begin_. Your Apple Watch will begin tracking it immediately. During tracking, your Apple Watch will display useful information to you depending on the workout. This can include the current duration, your current heart rate, distance you have travelled and more. (Figure 12.2)

67

Chapter 12 | Fitness & Exercise

Duration

Calories

Current heart rate

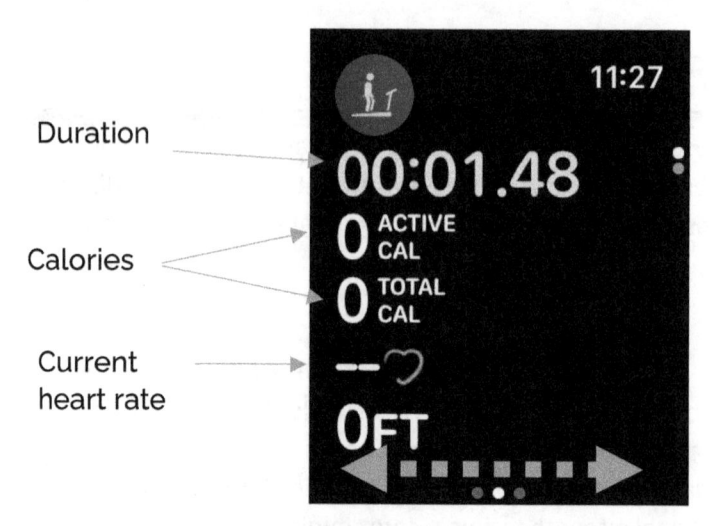

Figure 12.2 – Workout Display

To end a workout, simply swipe to the left and tap End. (Figure 12.3)

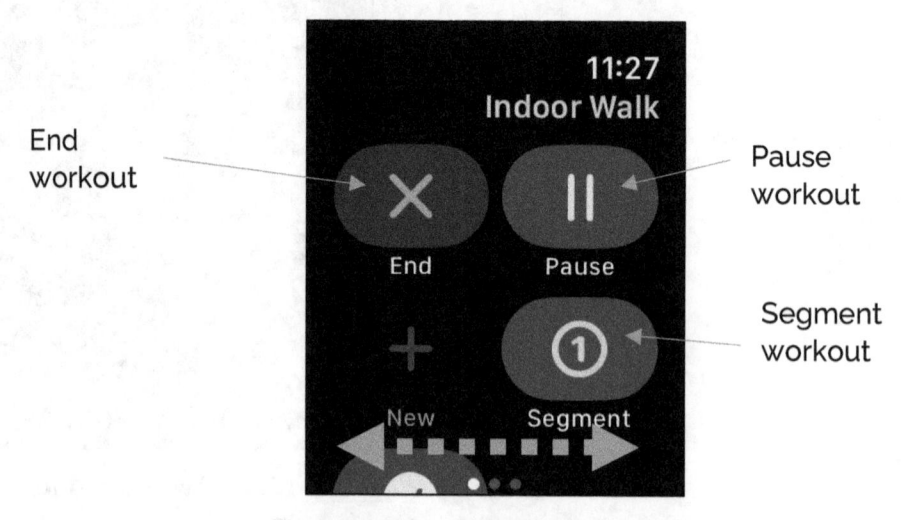

End workout

Pause workout

Segment workout

Figure 12.3 – Ending a Workout

All workout data tracked by your Apple Watch will integrate with the health features of your device.

68

Chapter 13 | Health Features

Chapter 13 – Health Features

There are many health features built in to the Apple Watch and they all work seamlessly together. In this detailed chapter, we will explore many of these features and how to utilize them.

Health features can be accessed on your Apple Watch through various apps. We went over two of these apps already, the Activity app and the Workout app. All the health apps on your Apple Watch feed into the main Health app, which stores and tracks all your health data on your iPhone.

Mindfulness App

The Mindfulness app is an app for mental well-being and offers exercises to improve your mental health. (Figure 13.1)

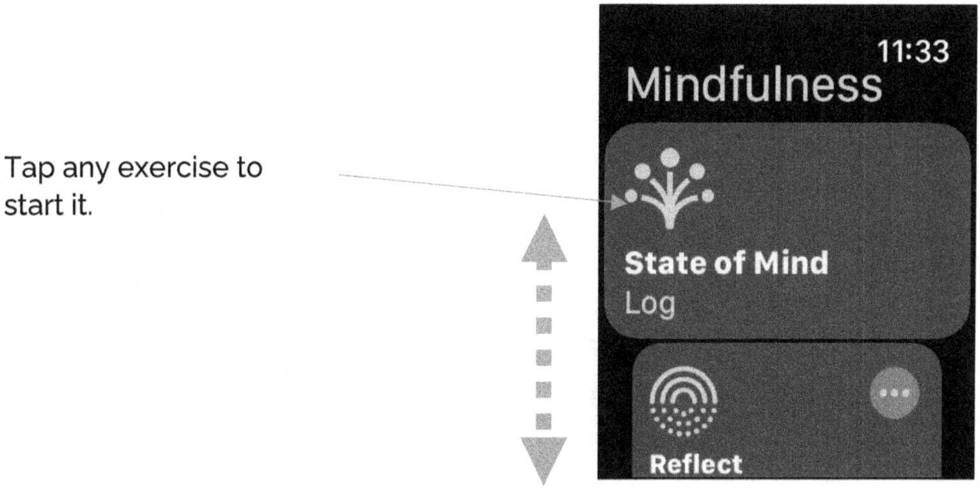

Tap any exercise to start it.

Figure 13.1 – Mindfulness App

State of Mind
The State of Mind feature allows you to log how you are currently feeling and to see how your feelings change over time.

69

Chapter 13 | Health Features

Reflect

The Reflect feature offers a quick, simple, reflection that you can think about and carry with you. It provides an opportunity to relax and deconstruct your thoughts.

Breathe

The Breathe feature will guide you through deep breathing exercises.

I recommend you give the Mindfulness app a try and then try the different exercises available within them. Each time you do one, it will be tracked in the Health app.

Sleep App

The Sleep app (Figure 13.2) tracks your sleep and encourages you to follow good sleeping hygiene practices. Your Apple Watch can automatically determine when you are sleeping and will track various metrics using its sensors including heart rate, blood oxygen, and sleep cycle.

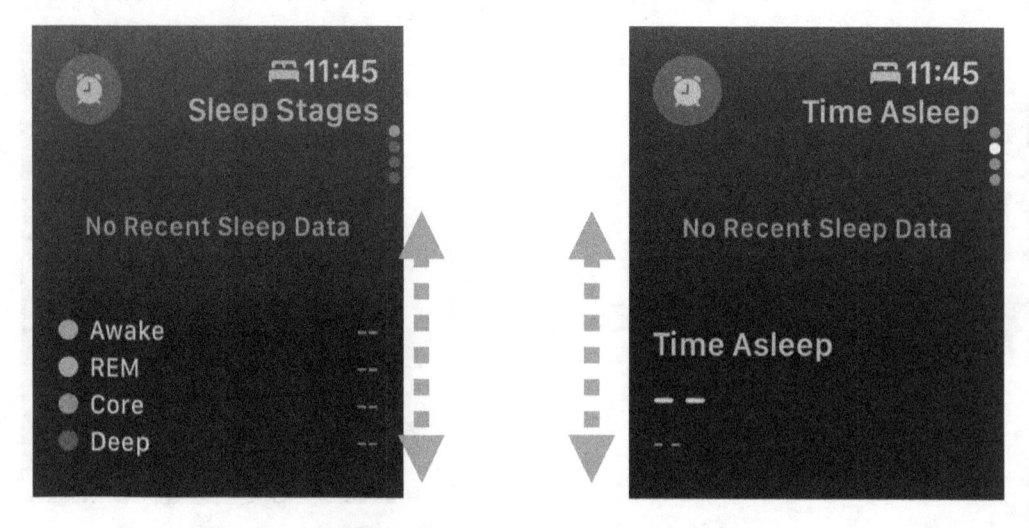

Figure 13.2 – Sleep App

Before you can start tracking your sleep, it is worth setting up your Bedtime routine **on your iPhone:**

1. Open the Health app **on your iPhone**.
2. Tap Sleep. (Figure 13.3)

Chapter 13 | Health Features

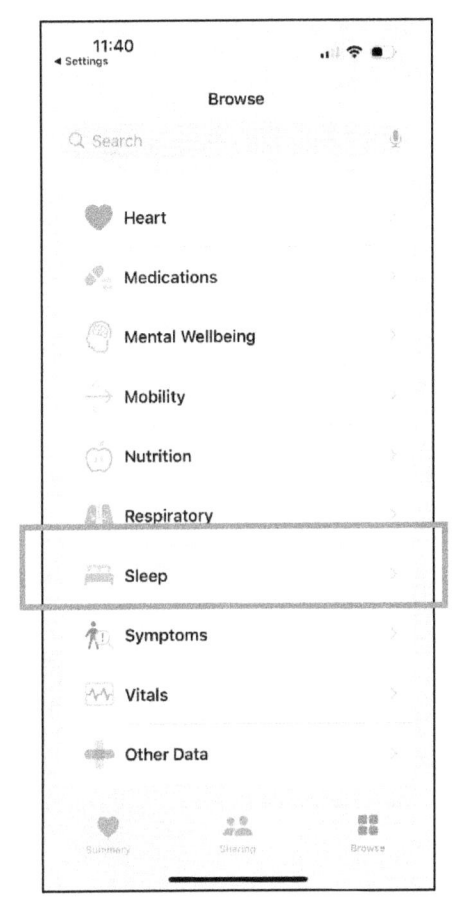

Figure 13.3 – Health App -> Browse Tab

3. Scroll down to Your Full Schedule and tap <u>Edit</u>.
4. On the next screen, set your daily ideal sleep schedule. (<u>Figure 13.4</u>)
    a. Be sure to select all the days your schedule applies to.
    b. Set your ideal bedtime and wakeup time.
    c. By scrolling down, you can set your alarm settings.

71

Chapter 13 | Health Features

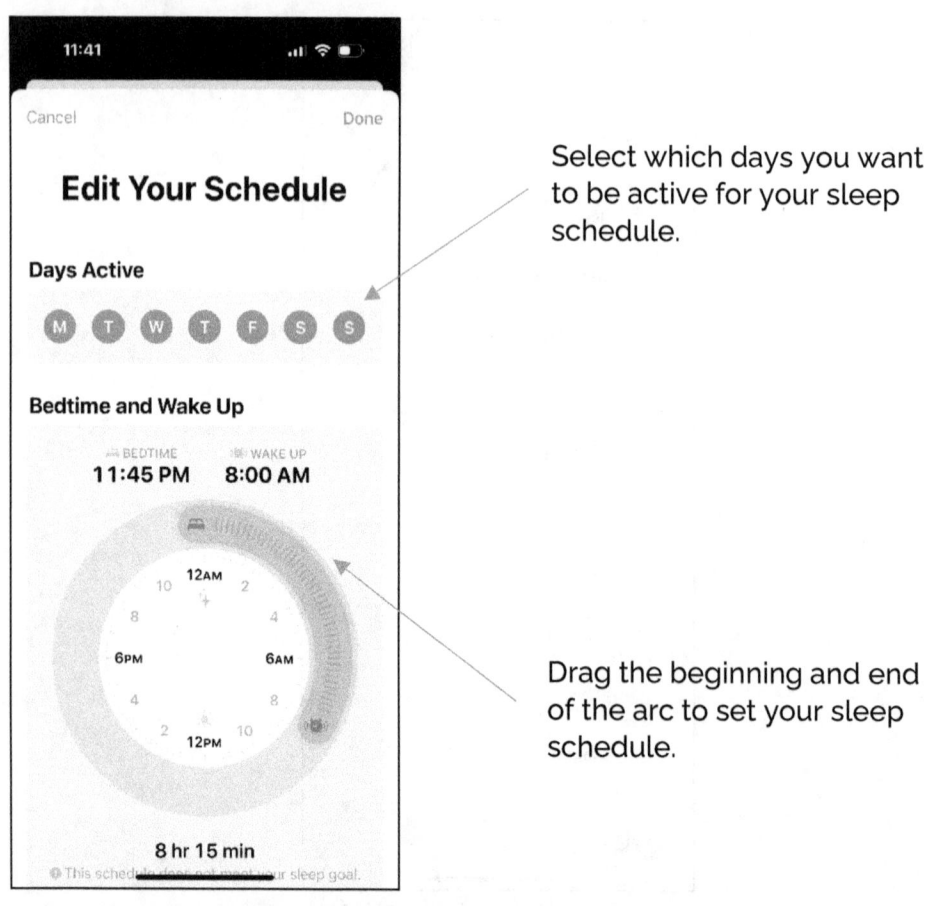

Select which days you want to be active for your sleep schedule.

Drag the beginning and end of the arc to set your sleep schedule.

Figure 13.4 – Health App -> Sleep -> Full Schedule

5. When finished here, return to the Schedule screen on your iPhone and set your Sleep Goal and Wind Down time.
6. Lastly, tap Manage Sleep in the Apple Watch App and make sure Track Sleep with Apple Watch is enabled. (Figure 13.5)

72

Chapter 13 | Health Features

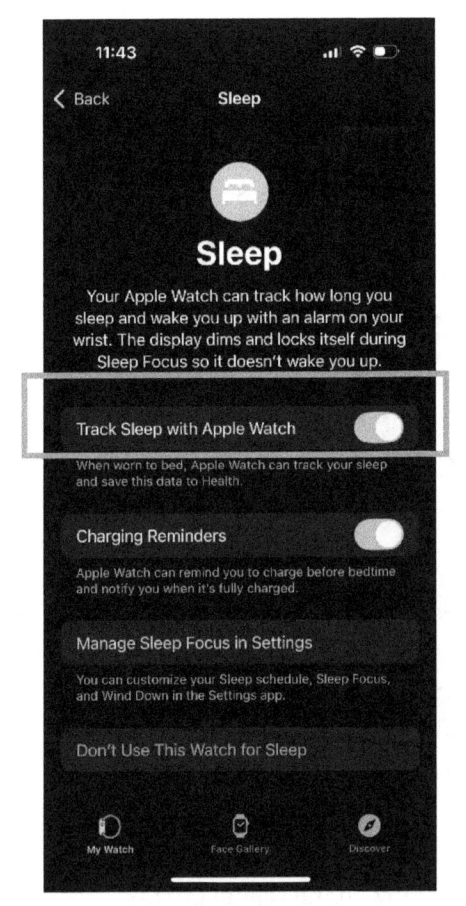

Figure 13.5 – Watch App -> Sleep

Now that your sleep schedule is set, your Apple Watch will notify you when it is time to start winding down for sleep. A wakeup alarm will also be automatically set up for each day on your schedule. You can easily cancel or edit your daily alarm in the Clock app in the Alarms tab on your iPhone.

For the Sleep app to track your sleep, you need to wear your Apple Watch while sleeping.

ECG App
Newer Apple Watch models can take an ECG, which is an electrocardiogram. What this does is record the timing and strength of the electrical signals of your heart. Your

Chapter 13 | Health Features

watch can also determine if the ECG detects an atrial fibrillation and allows you to share the results of the ECG with your doctor.

To use this feature, follow these steps:

1. Open the ECG app from the Apps Screen.
2. Before processing, make sure your Apple Watch is snug on your wrist and then follow the instructions on your screen. (Figure 13.6)

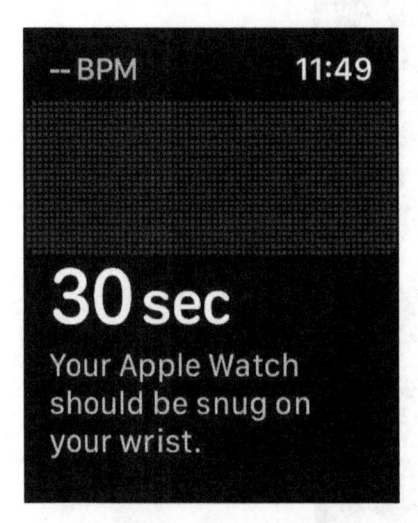

Hold your index finger lightly on the face of the Digital Crown and hold it there until the timer elapses to complete the ECG.

Figure 13.6 – ECG App

3. Hold your finger lightly on the Digital Crown.
4. Continue holding your finger lightly on the Digital Crown for 30 seconds as the timer counts down. When complete, your results will be shown to you on your Apple Watch. You can view a full report of the ECG in the Health app **on your iPhone**. (Figure 13.7)

Chapter 13 | Health Features

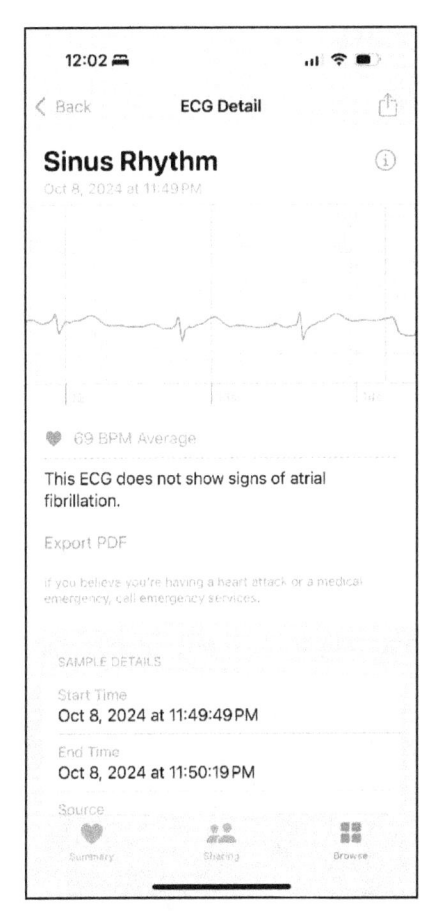

Figure 13.7 – ECG Results in Health App (iPhone)

Heart App

The Heart app measures and displays data about your heart. To get started, open the Heart app from the Apps Screen.

The Heart app has 4 screens. The first screen actively takes your current heart rate. (Figure 13.8)

Chapter 13 | Health Features

Current heart rate

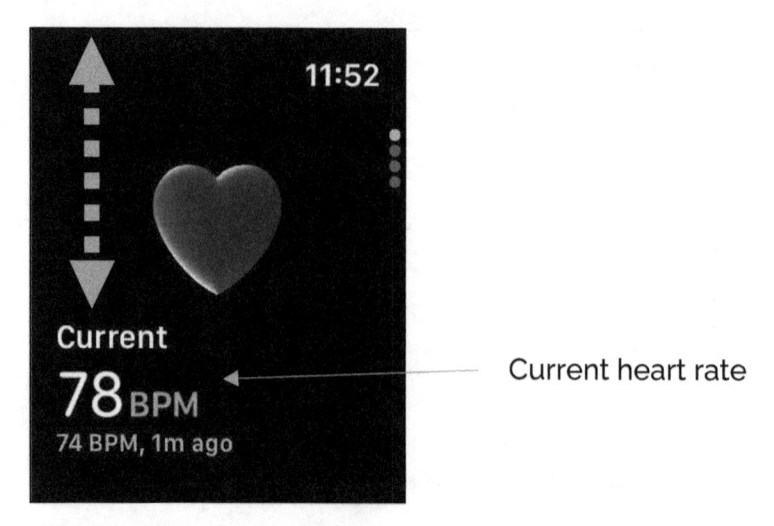

Figure 13.8 – Heart App

Swipe down to your daily heart rate range. Your Apple Watch will automatically record your heart rate periodically throughout the day. (Figure 13.9)

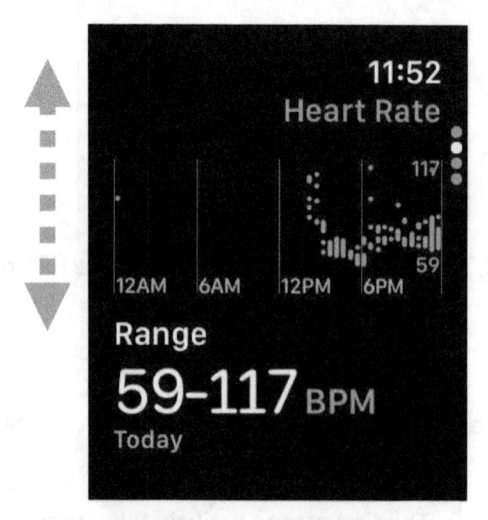

Figure 13.9 – Heart Rate Range

Swipe down again to see your average resting heart rate for the day. (Figure 13.10)

Chapter 13 | Health Features

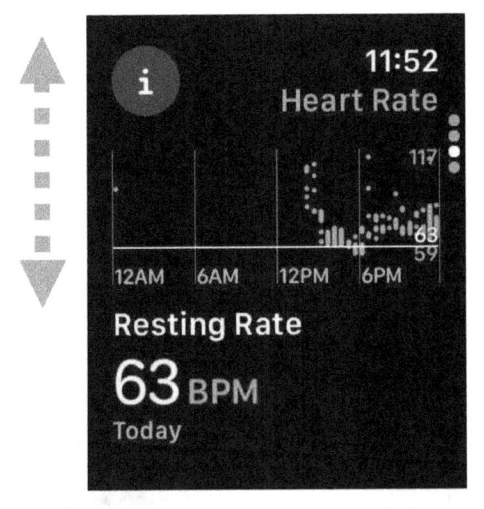

Figure 13.10 – Resting Heart Rate

Swipe down one more time to see your walking average heart rate for the day. (Figure 13.11)

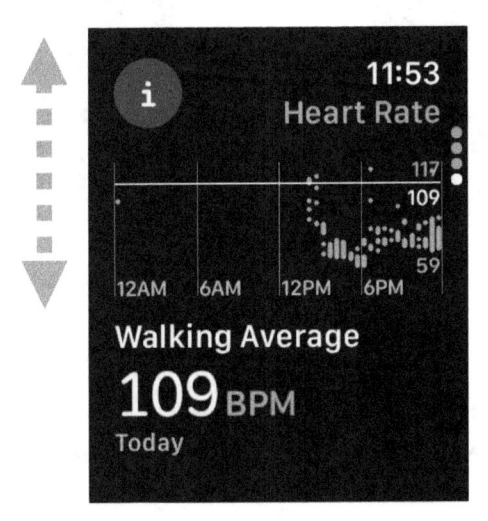

Figure 13.11 – Walking Average Heart Rate

You can set up notifications for the Heart app from the Watch app **on your iPhone** by tapping on Heart from the My Watch tab. (Figure 13.12)

77

Chapter 13 | Health Features

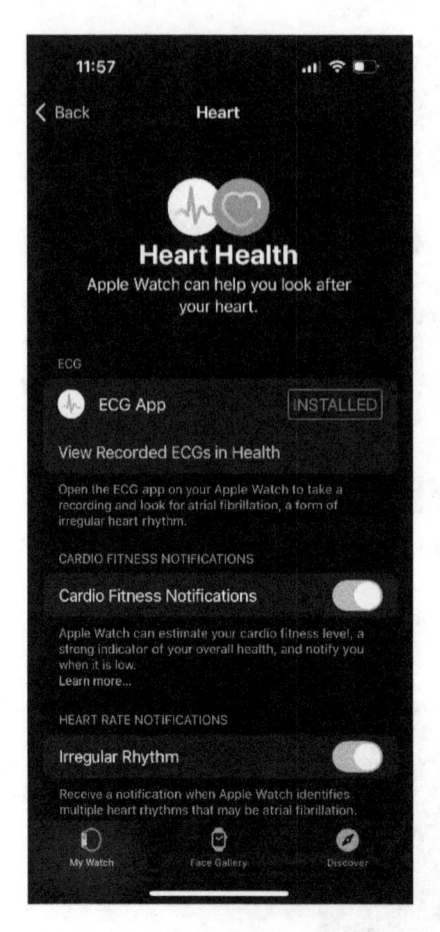

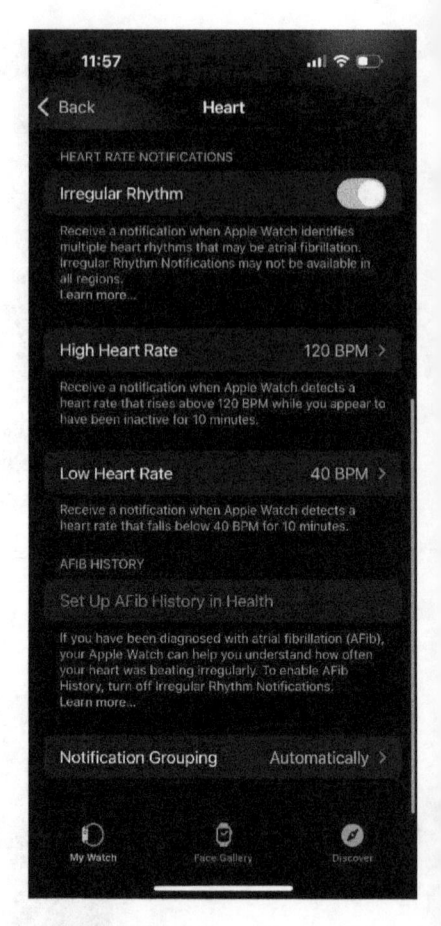

Figure 13.12 – Heart Rate Notifications

Noise App

The Noise app monitors the sound levels you experience throughout the day and determines if you have been exposed to excessive noise levels. While using the Noise app, you can see the current sound level of your environment in decibels. (Figure 13.13)

Chapter 13 | Health Features

This meter indicates the current sound level of your environment.

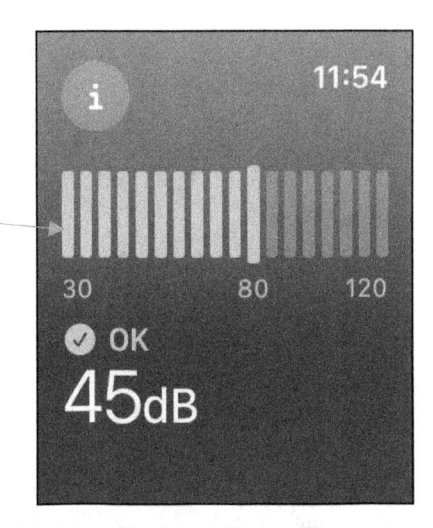

Figure 13.13 – Noise App

You can set up noise monitoring and notifications in the Watch app **on your iPhone** by tapping on Noise from the My Watch tab. (Figure 13.14)

Chapter 13 | **Health Features**

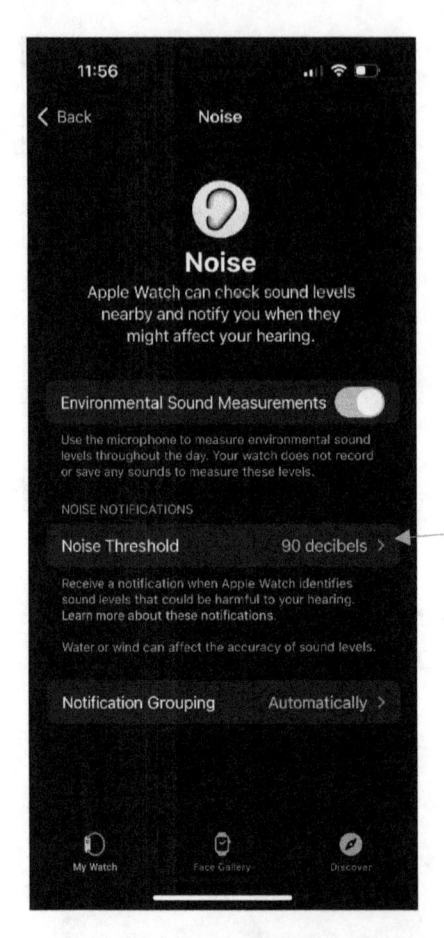

Set your decibel level and your Apple Watch will notify you when noise exceeds that level.

Figure 13.14 – Watch App -> My Watch -> Noise

Cycle Tracking

You can track your cycle and gain valuable insights about it using the Cycle Tracking app on the Apple Watch. Before getting started using this app, you must set up Cycle Tracking in the Health app **on your iPhone.**

1. Open the Health app on your iPhone.
2. Tap the Browse tab at the bottom.
3. Tap Cycle Tracking under Health Categories.
4. Click Get Started.
5. Follow the instructions on your screen to set up Cycle Tracking.

Chapter 13 | Health Features

Once Cycle Tracking is set up. Your Apple Watch will notify you with important updates and you can quickly view your cycle's details inside the Cycle Tracking app.

Vitals App

The Vitals app tracks your key health metrics measured during sleep to provide you insights about your overall health. To get started using this feature, open the Vitals app from the Apps Screen on your Apple Watch. Follow the instructions on your screen and the app will tell you it needs seven days of sleep data to provide you with insights. All you need to do is wear your Apple Watch while you go to sleep for seven days and then the Vitals app will start delivering notifications to you about your health. It will also look for significant changes in your key health metrics and notify you when it encounters them.

Health App

As mentioned earlier in this chapter, all the health apps and health features of your Apple Watch feed into the Health app. The Health app is best accessed **on your iPhone** and provides you with a single place to access all your health data. This app is huge and has endless features, so we will not cover it in its entirety here, but I will show you how to access some useful features inside it.

Viewing your Health Summary

In the Summary tab of the Health app **on your iPhone**, you can view various data about your health tracking. This page can include your average steps per day, any noise notifications, and much more. (Figure 13.15)

Chapter 13 | Health Features

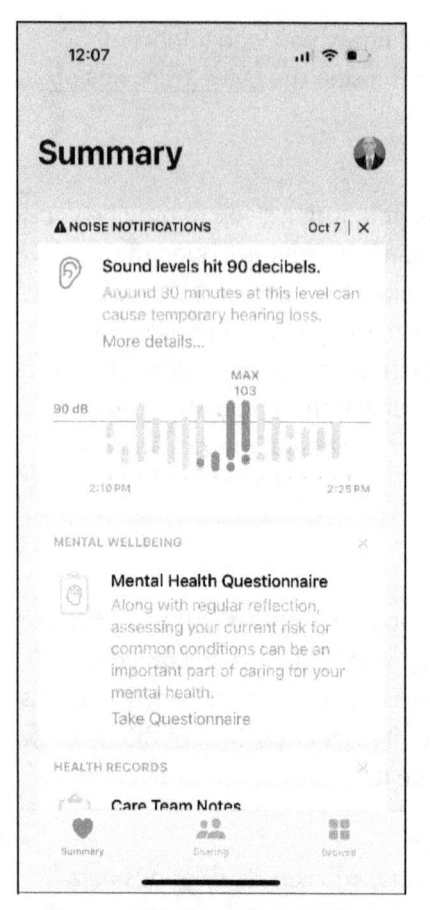

The Summary tab shows you recent health updates and recommendations.

Figure 13.15 – Health App -> Summary

Viewing your Heart Data

To view your heart data in the Health app, tap the Browse tab and then tap Heart. From here you can see various metrics about your heart, including your heart rate variability, latest heart rate measurement, ECGs, and more. You can tap on any metric to view more information. (Figure 13.16)

Chapter 13 | Health Features

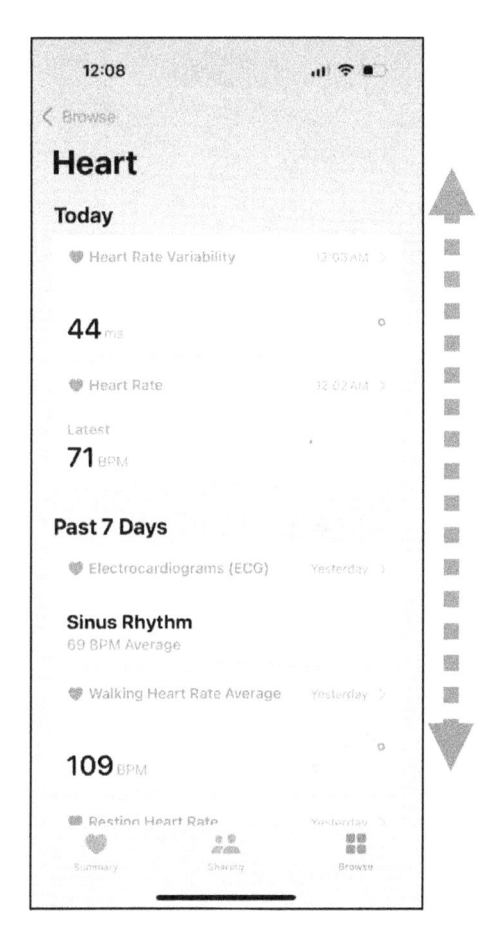

[screenshot of Health app Heart screen showing Today: Heart Rate Variability 44 ms, Heart Rate Latest 71 BPM; Past 7 Days: Electrocardiograms (ECG) Sinus Rhythm 69 BPM Average, Walking Heart Rate Average 109 BPM, Resting Heart Rate]

Figure 13.16 – Health App -> Browse Tab -> Heart

Viewing your Sleep Data

To view your sleep data in the Health app, tap the Sleep option in the Browse tab. From here, if you have been tracking your sleep with your Apple Watch, you can see your sleep statistics.

Viewing your Activity & Mobility Data

Your Apple Watch tracks so much about your movement and organizes it nicely in the Health app. To see your movement activity, tap the Activity option. From here, you can see the steps you took, calories, burned, and more. (Figure 13.17)

83

Chapter 13 | **Health Features**

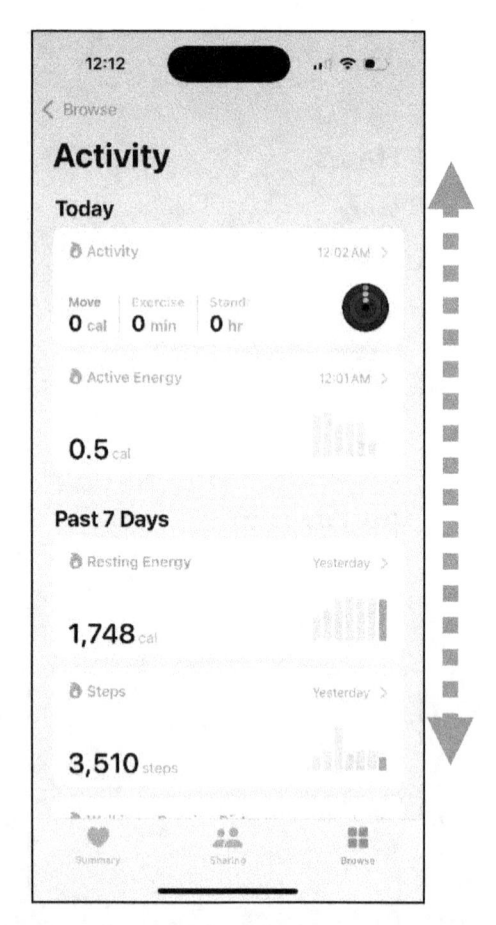

Figure 13.17 – Health App -> Browse Tab -> Activity

To see your mobility data, tap the Mobility option in the Browse tab. This screen will show you some advanced mobility data that you may find useful.

Chapter 14 – Safety Features

The Apple Watch has many safety features that can be very beneficial. In this chapter, we will cover these features and how to set them up properly.

Medical ID

The Medical ID feature allows you to show important medical information about yourself on your Apple Watch for anyone to see. In an emergency, anyone attending to you can quickly pull up your Medical ID and see the information you want to share, such as your age, medical conditions, blood type, and allergies.

To set up your Medical ID (Figure 14.1), you will need to **use your iPhone:**

1. Open the Health app.
2. Tap your *image* at the upper right and then tap Medical ID.
3. Click Get Started or Edit, and then enter your information. You can enter various information including your emergency contacts.
4. The top two options allow you to control when your Medical ID can be shared.

NOTE: Anything you put in your Medical ID can potentially be seen by anyone who has your iPhone or Apple Watch.

Chapter 14 | Safety Features

When enabled, anyone can view your Medical ID by looking at your iPhone or Apple Watch in an emergency.

When enabled, your iPhone or Apple Watch will share your Medical ID with emergency services if they are called.

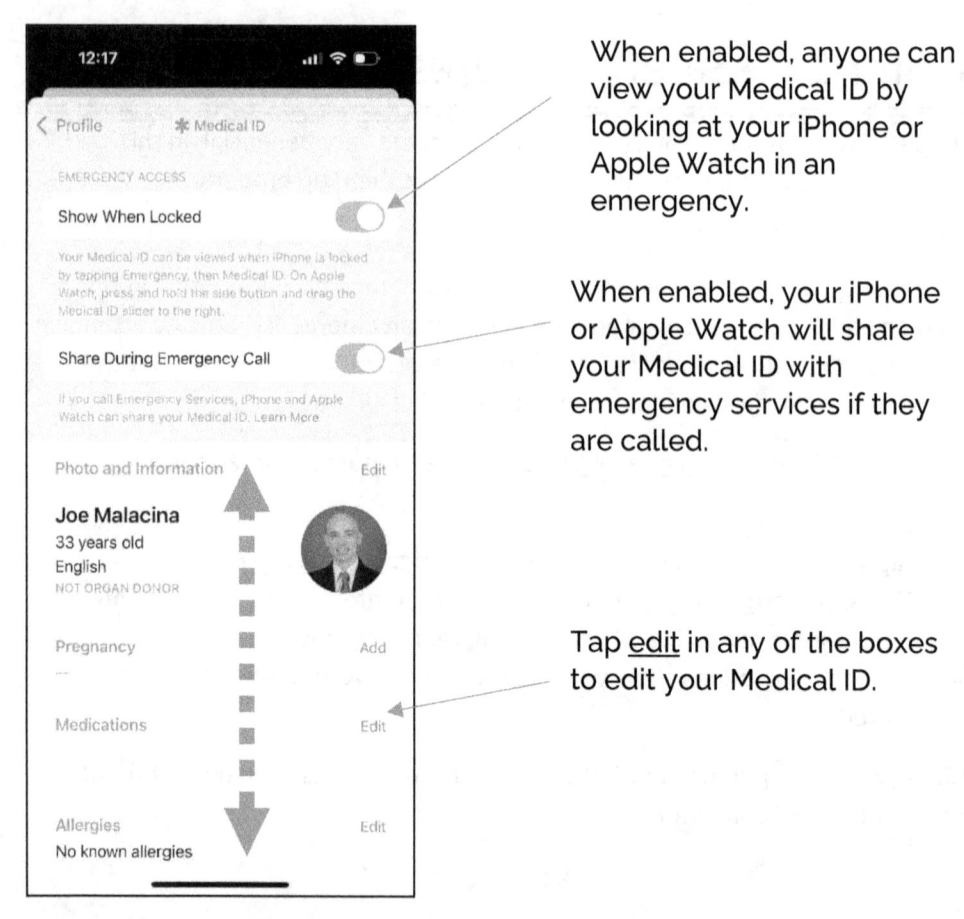

Tap edit in any of the boxes to edit your Medical ID.

Figure 14.1 – Medical ID

Contact Emergency Services

You can contact Emergency Services quickly from your Apple Watch if the need arises. There are three ways to use this feature:

Method 1: Tap and hold the side button on your Apple Watch until the sliders appear (Figure 14.2). Tap and slide the SOS icon to the right over emergency call. This will call "911" or your local emergency number.

Chapter 14 | Safety Features

Tap and slide this icon to the right to view your Medical ID.

Tap and slide the SOS icon to the right to call emergency services.

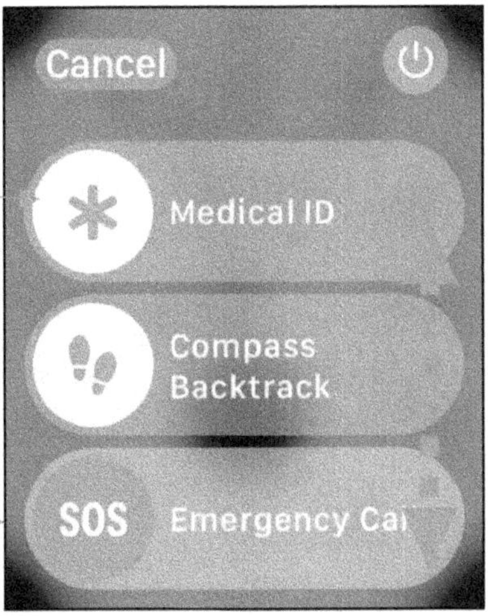

Figure 14.2 – Emergency Services

Method 2: Tap and hold the side button and keep holding it until your Apple Watch starts playing a warning sound and starts counting down. When the countdown ends, your watch will automatically call emergency services.

Method 3: Say "Hey Siri, call 911".

Fall Detection
Your Apple Watch can alert emergency services along with your emergency contacts when it detects that you have taken a hard fall. Before calling emergency services, it will first notify you. If you do not respond to the notification, it will make the call. To use this feature, first make sure you have emergency contacts set up in your Medical ID. Fall detection will be automatically enabled in your 55 or older. To set up fall detection (Figure 14.3):

1. Open the Settings app from the Apps Screen **on your Apple Watch**.
2. Tap SOS and then tap Fall Detection.
3. Tap the top slider to enable.
4. Set which option you would like (always on or on during workouts only).

Chapter 14 | Safety Features

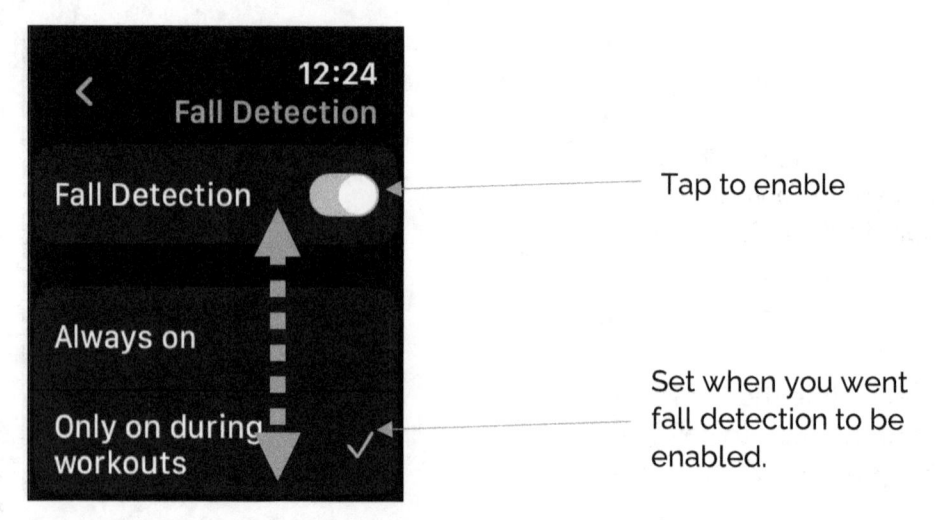

Figure 14.3 – Settings App -> SOS -> Fall Detection

Siren

Apple Watch Ultra models have a built-in siren that emits a loud noise to try and attract help and deter danger. To activate the siren on these models, press and hold the action button until the countdown finished. To stop the siren, tap the stop icon.

Crash Detection

Your Apple Watch can detect when a severe car crash has been detected and notify your emergency contacts along with emergency services. To set this up:

1. Open the Settings app from the Apps Screen **on your Apple Watch**.
2. Tap SOS and then make sure Call After Severe Crash is enabled.

If your Apple Watch ever detects a severe car crash, it will notify you on your watch. If you do not respond, your watch will call emergency services automatically and leave them with an audio message informing them a severe car crash has been detected. It will also tell emergency services your latitude and longitude location. In addition, it will text your emergency contacts the same information.

Chapter 15 | More Useful Apps & Features

Chapter 15 – More Useful Apps & Features

We have covered most of the core features of the Apple Watch, however there is still much to cover that you may find helpful. In this chapter, we will cover some additional apps and features that are included with your Apple Watch.

Mail App

Inside the Mail app on your Apple Watch (Figure 15.1), you can access all your email accounts that you use on your iPhone's mail app. The Apple Watch is not particularly great at composing new emails, but it is quite useful for quickly reading emails.

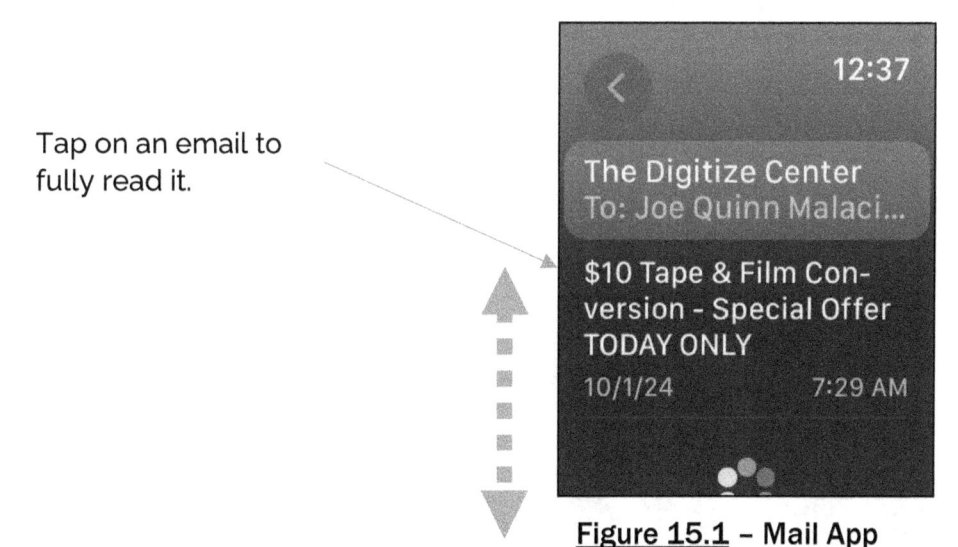

Tap on an email to fully read it.

Figure 15.1 – Mail App

Timer App

The Timer app is a useful app for quickly setting a timer on your Apple Watch (Figure 15.2). When the timer elapses, your watch will notify you with an alarm and a vibration.

Chapter 15 | More Useful Apps & Features

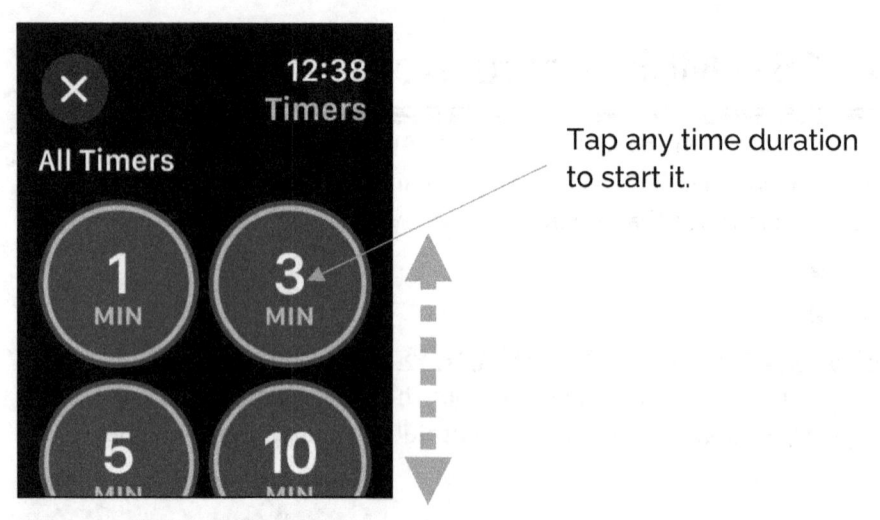

Figure 15.2 – Timer App

Tap any time duration to start it.

Stopwatch App

The Stopwatch app is a simple app that operates like an actual stopwatch. Tap the green icon to begin the stopwatch and tap the red icon to stop it. (Figure 15.3)

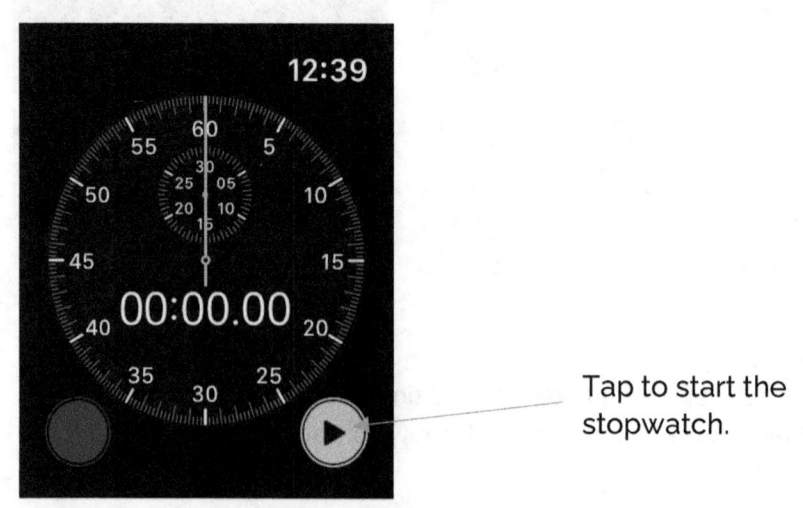

Tap to start the stopwatch.

Figure 15.3 – Stopwatch App

Chapter 15 | More Useful Apps & Features

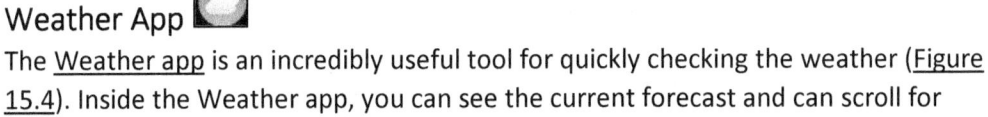

Weather App
The Weather app is an incredibly useful tool for quickly checking the weather (Figure 15.4). Inside the Weather app, you can see the current forecast and can scroll for more information. You can also tap on information to see different views, such as precipitation forecasts and air quality metrics.

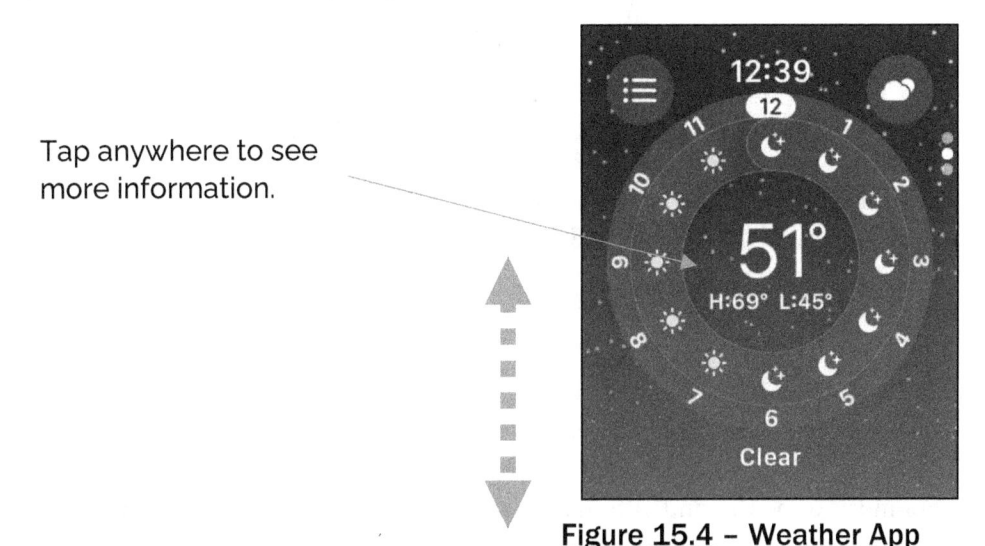

Tap anywhere to see more information.

Figure 15.4 – Weather App

Compass App
The Compass app is a nifty app that does exactly as it implies, acts as a compass. The compass app will show you your direction and elevation and updates in real-time. You can turn the Digital Crown to change your view.

91

Chapter 15 | More Useful Apps & Features

See details

Mark a waypoint

Record a path

Figure 15.5 – Compass App

Music App

The Music app allows you to browse through your music straight from your watch. If you are listening to music on your Apple AirPods, you can quickly browse through your music using this app. Alternatively, you can use the Now Playing app, which we will cover next.

Now Playing App

The Now Playing app is a very useful app that allows you to control whatever is currently playing on your iPhone, such as an audio track or a video. Whenever you are playing something on your iPhone, controls for this will appear on your Apple Watch in what is called "Now Playing". To access Now Playing, you can find it in your Smart Stack (turn the Digital Crown upwards) or by tapping the Now Playing icon at the top of your home face (Figure 15.6).

Chapter 15 | More Useful Apps & Features

Tap the Now Playing icon at the top of your home face to quickly open it.

Figure 15.6 – Access Now Playing

While in Now Playing, you can skip to the next track or video, go back, or even adjust the volume of the audio. To adjust the volume, simply turn the Digital Crown upwards or downwards. (Figure 15.7)

Previous track

Next track

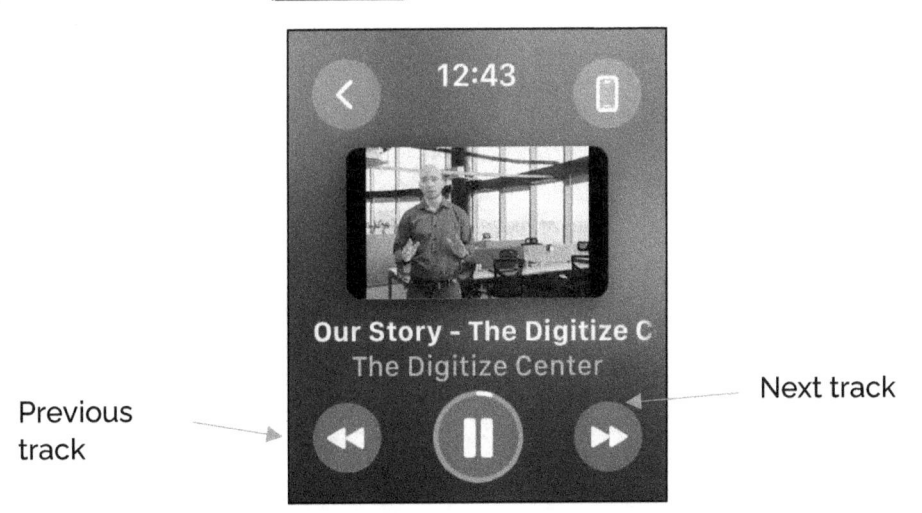

Figure 15.7 – Now Playing

The Now Playing feature is a great tool for quickly controlling what is currently playing on your iPhone.

93

Chapter 15 | More Useful Apps & Features

Downloading New Apps

Many third-party apps not included with your iPhone can be loaded onto your Apple Watch. Whenever you download an app to your iPhone, if that app has an Apple Watch version, it will automatically be downloaded onto your watch as well. You can access these apps from the Apps Screen and interact with them the same way you can the default apps. In addition, any apps that deliver notifications to you on your iPhone will also deliver notifications to your Apple Watch, even if it does not have an Apple Watch app. If you do not want an app to deliver notifications to your Apple Watch, you can manage this in the Apple Watch app inside Notifications **on your iPhone**.

To download new apps to your iPhone, use the App Store app.

Chapter 16 | Apple Pay

Chapter 16 – Apple Pay

Apple Pay allows you to store your credit and debit cards digitally on your iPhone and Apple Watch. You can then pay using either of these devices if the merchant accepts Apple Pay. This feature is incredibly convenient, and I highly recommend giving it a try.

Setting Up Apple Pay

It is recommended that you set up Apple Pay completely **on your iPhone**, as it is much easier to do so.

1. **On your iPhone** open the Wallet app (Figure 16.1).

Add a new card to your Wallet.

View your cards and passes.

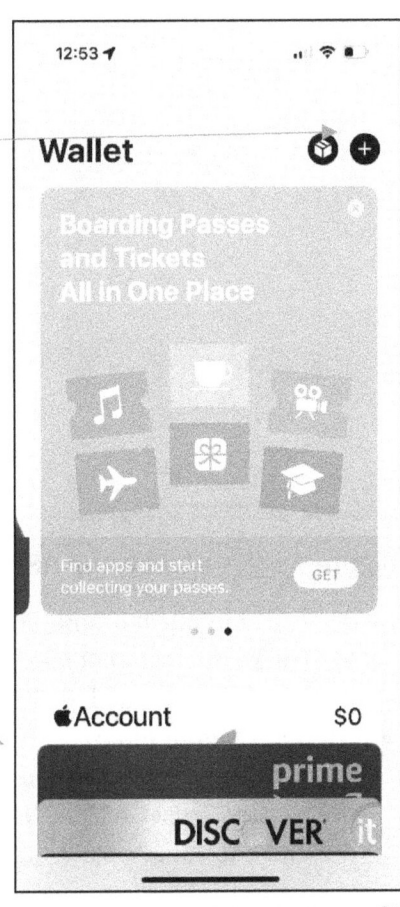

Figure 16.1 – Wallet App (iPhone)   95

Chapter 16 | Apple Pay

2. Tap the plus sign in the upper right.
3. Tap the choice that you want to add. You can add various types of cards including credits cards, transit cards, and IDs. (Figure 16.2)

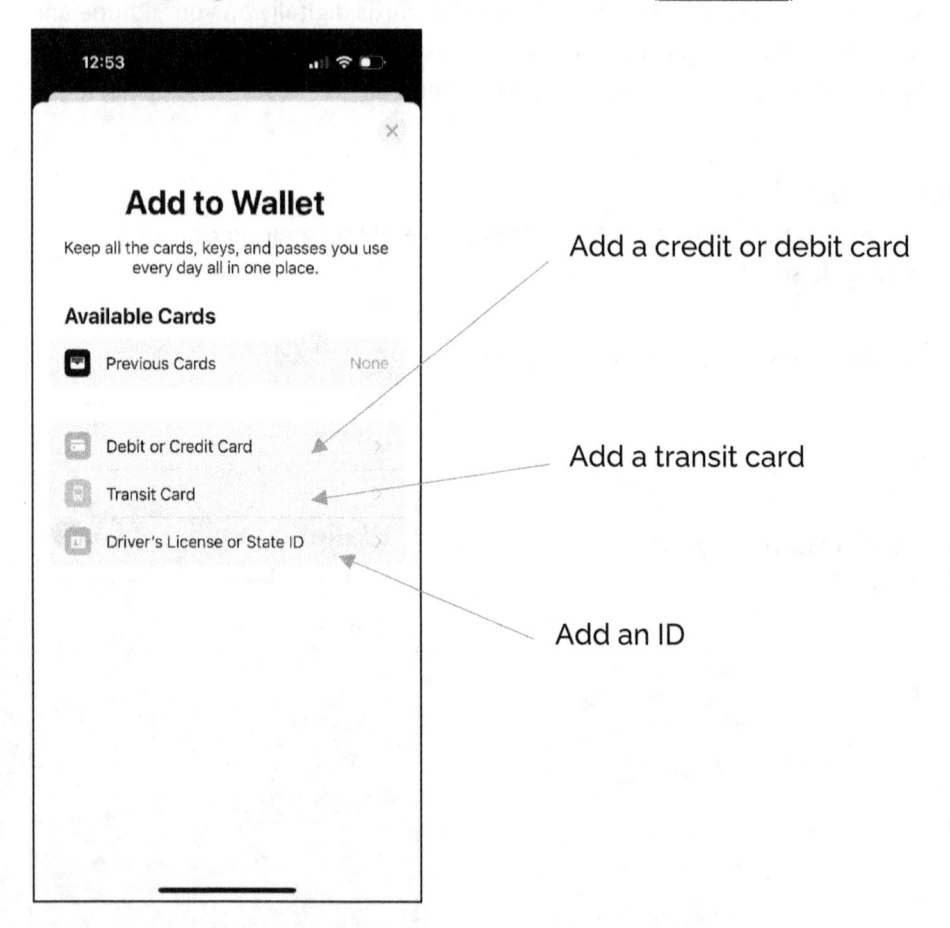

Add a credit or debit card

Add a transit card

Add an ID

Figure 16.2 – Wallet App (iPhone) -> Add New Card

4. Tap Continue if prompted.
5. Follow the instructions on your screen to add your card(s).
6. Once you have added all the cards you want to your iPhone, open the Watch app on your iPhone.
7. In the My Watch tab, tap the Wallet & Apple Pay option.
8. Under OTHER CARDS ON YOUR PHONE, you can tap ADD next to each one to add it to your Apple Watch. (Figure 16.3)

Chapter 16 | Apple Pay

Tap ADD next to each card to add it to your Apple Watch.

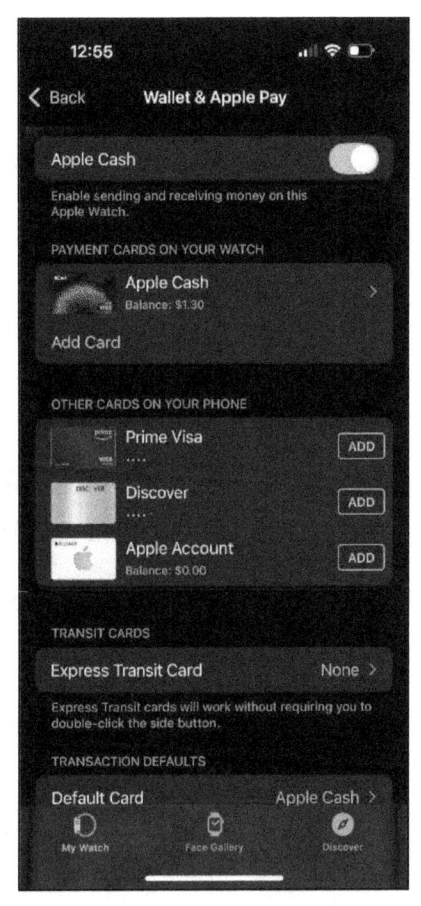

Figure 16.3 – Watch App -> Wallet

9. You may need to enter some additional information to add these cards to your watch, such as the CVV codes of the cards.

Using Apple Pay
To use Apple Pay on your watch, follow these steps:

1. When ready to pay at an Apple Pay enabled terminal, **double press** the side button on your Apple Watch.
2. Swipe to the card you want to use. (Figure 16.4)

97

Chapter 16 | Apple Pay

Figure 16.4 – Using Apple Pay

3. Hold your Apple Watch near the payment terminal until you are notified of the transaction.

Chapter 17 – Siri

Siri, as you may have heard, is Apple's intelligent voice assistant. You can talk to Siri, and Siri will listen to what you say and try to do whatever you ask. The possibilities of what you can do with Siri are endless.

Accessing Siri
Siri can be accessed in 3 different ways on your Apple Watch.

- Say "Siri" or "Hey Siri".
- Raise your Apple Watch to your mouth and speak to it.
- Hold the Digital Crown.

Any of the above 3 methods will work for accessing Siri. You can alter these settings in the Watch app **on your iPhone** if you would like.

Using Siri
So, what can we use Siri for? Siri is constantly evolving and improving over time. If you have a new iPhone model, then Siri is extremely powerful with its Apple Intelligence capabilities. Regardless, you can say pretty much anything to Siri and then see how she responds. Depending upon your request, she can take direct action or bring up additional information to help you. Here are some examples:

- "Call John Doe" – Siri will call John Doe from my contact list.
- "Set a timer for 5 minutes" – Siri will set a timer for 5 minutes and start it.
- "Change my wake-up alarm to 7 AM tomorrow" – Siri will change your wake-up alarm to 7 AM for tomorrow only.
- "Send a text message to Jane Doe saying I am running late." – Siri will send a text message to Jane Doe from your contact list with a message of *I am running late*. Siri will confirm this with you before sending.
- "What is 2+2?" – Siri will do the math and tell you the answer is 4.
- "Turn off the lights." – Siri will turn off the lights that are connected to the Home app on your iPhone.
- "What time is it?" – Siri will speak the current time to you.
- "What is the weather tomorrow?" – Siri will speak and display the weather forecast for tomorrow at your location.

Chapter 17 | Siri

Chapter 18 | Tips & Tricks

Chapter 18 – Tips & Tricks

Congratulations! You have made it through most of this book, and you should now have a solid understanding of how to use your Apple Watch. There are no more "basics" left to teach you, so I will leave you with a few tips and tricks that you may find helpful when using your Apple Watch.

Background Apps & the App Switcher

(See Figure 18.1) Whenever you open an app, it remains open in the background even after you leave the app. In some circumstances, it can be beneficial to completely close an app running in the background if it is not working properly. You can also use this method to quickly go to an app you were recently using.

To do this:

1. Quickly double press the Digital Crown.
2. This will bring up the App Switcher. Here, you can swipe up and down through your recently used apps. To go to one, simply tap on its screen. To completely close a recently used app, tap and swipe it to the left, and then tap the x.

Tap on any app screen to go directly to that app.

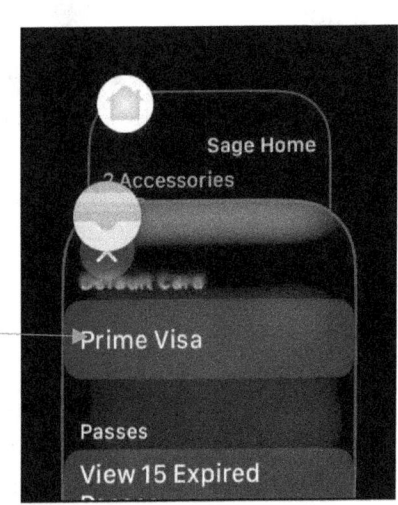

Figure 18.1 – App Switcher

101

Chapter 18 | Tips & Tricks

Low Power Mode

If your Apple Watch's battery is running low and you need to extend it until you have the opportunity to charge, you can enable low power mode (Figure 18.2). With low power mode enabled, certain features of your watch will be disabled to save power including the always-on display, some sensors, and notification fetching. To turn on low power mode:

1. Open the Control Center by pressing the side button.
2. Tap the battery percentage icon.
3. Tap Low Power Mode.

Close

Enable Low Power Mode

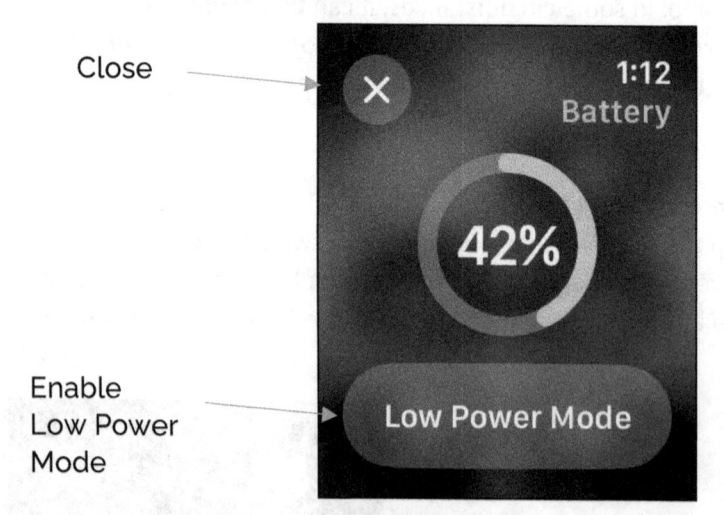

Figure 18.2 – Low Power Mode

Low power mode can be turned off in the same way.

Smart Stack

The Smart Stack is a set of widgets that will display on your Apple Watch (Figure 18.3). This Smart Stack can be accessed from your home face by turning the digital crown upwards. You can scroll through this stack of widgets to see information that may be relevant to you. You can tap on any of the widgets to see more information or to be brought to its associated app. The Smart Stack can be customized from within the Watch app **on your iPhone**.

Chapter 18 | Tips & Tricks

Tap on any widget to see more information.

Figure 18.3 – Smart Stack

The Double Tap Gesture
The Double Tap Gesture is a unique feature that allows you to control your Apple Watch without even touching it. What this gesture does is perform a default action depending on what is happening with your Apple Watch. In other words, it will do something different depending upon what your Apple Watch is doing.

To perform the double tap gesture, simply tap your index finger to your thumb twice consecutively on your watch hand. The gesture will do the following on your Apple Watch depending on the circumstance:

- Home Face – will open your Smart Stack.
- Incoming phone call – will answer the call.
- During a timer countdown – will pause, resume, or end a timer.
- An alarm clock is alerting – will snooze the alarm.
- Now playing is active – will pause or play the media.
- Camera app is open – will capture the photo.
- A notification is being viewed – will perform the primary notification action.

Using the Camera App
A very helpful tip that you can do on your Apple Watch is control your iPhone's camera. How many times have you been in a situation where you want to take a

103

Chapter 18 | Tips & Tricks

group photo but do not want to exclude the person who must capture the photo? Well, with the Apple Watch, that problem is solved. You can set up your iPhone to take a photo, and then actually take the photo using your Apple Watch. Here's how:

1. Set up your iPhone in a location to take a photo. It is recommended that you use the selfie lens so you can see a preview of the photo.
2. Open the Camera app from the Apps Screen **on your Apple Watch**.
3. Tap the three small dots to enable a 3 second timer. (Figure 18.4)

Go back to camera

Enable a 3 second timer

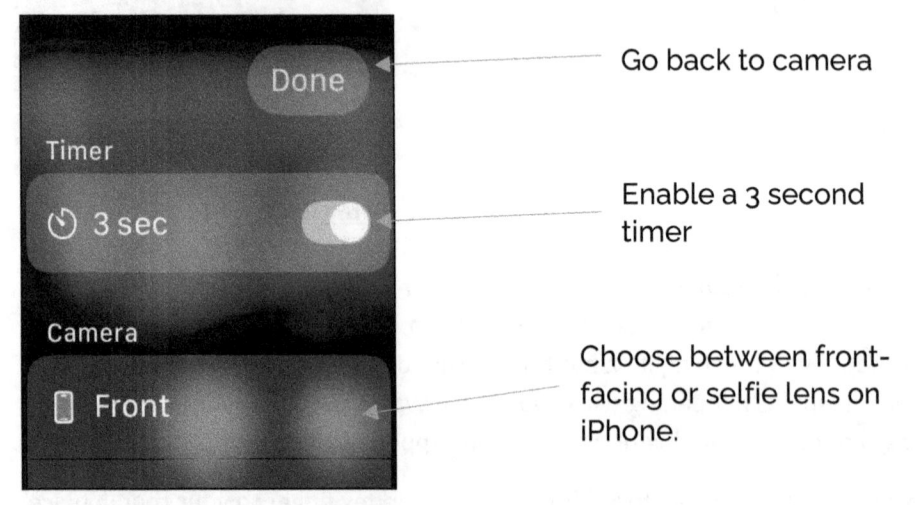

Choose between front-facing or selfie lens on iPhone.

Figure 18.4 – Camera App Settings

4. When ready to take the photo, tap the capture icon **on your Apple Watch**.
5. A three second timer will elapse, allowing you to prepare for the photo, and then your iPhone will capture the photo. (Figure 18.5)
6. You can preview the photo on your Apple Watch using the image at the bottom left.

Chapter 18 | Tips & Tricks

Options

Take photo
or start
timer.

Figure 18.5 –Camera App

Unlock your Apple Watch with your iPhone
You can use your iPhone to unlock your Apple Watch, rather than entering your passcode. This works by your iPhone detecting that your Apple Watch is on your wrist, and your iPhone will automatically unlock it when you go to use it. To enable this:

1. Open the Settings app from the Apps Screen **on your Apple Watch**.
2. Tap Passcode.
3. Make sure Unlock with iPhone is enabled.

Quickly Set a Photo as an Apple Watch Face
On your iPhone, you can quickly set a Photo as your Apple Watch Face:

1. Open the Photos app **on your iPhone**.
2. Find the photo you want to set as your face and tap it to bring it to full screen.
3. Tap the rectangle and arrow icon.
4. Tap Create Watch Face.

105

Chapter 18 | Tips & Tricks

View your Apps Screen as a List

You can view your Apps Screen as a List instead of a grid. This can be helpful if you have many apps installed and have trouble locating certain ones. To do this, simply scroll all the way down on the Apps Screen and tap List View. (Figure 18.6)

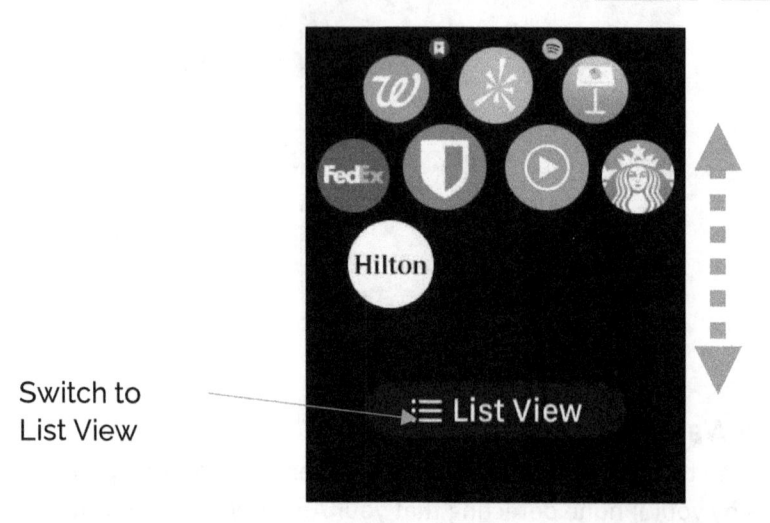

Switch to
List View

Figure 18.6 –Apps Screen List View

106

Chapter 19 – Conclusion & More Resources

Thank you for taking the time to read this book. It is my hope that you feel much better and confident about using your Apple Watch. I am personally confident that if you took the time to read this entire book, then you will have no problem using every aspect of your Apple Watch with ease. Continue to use this book as a reference when you need it. The table of contents can quickly lead you to your answer and the index can help you as well.

I welcome your thoughts and feedback on this text, please come visit my Facebook page online at facebook.com/joemalacina1. You can also tweet me @JoeMalacina on Twitter/X. I am often online answering questions from people who have read this text and helping people with their Apple device issues. As a last piece of advice, please remember your lock screen passcode. Forgetting this passcode can lead to a real headache.

Enjoy using your Apple Watch!

More Resources – Infinity Guides

This guide has covered all the beginner aspects of the Apple Watch. We have also covered many intermediate and advanced aspects, but there is still plenty more you can learn. Most notably, there is a lot to learn about using your iPhone in conjunction with your Apple Watch. You can also learn more about Apple Music, iTunes, and social media on your iPhone.

Infinity Guides is an excellent resource for beginners, and on the Infinity Guides website you can find books, manuals, DVDs, and online courses made for beginners. The online courses can be especially helpful as most of them are about 30 minutes long and teach you through video instruction. I have created a short list of things you can learn with Infinity Guides for your reference.

www.infinityguides.com **Content:**

- *iPhone Manual for Beginners* - **RECOMMENDED**
- *Facebook for Beginners*
- *Twitter for Beginners*

Chapter 19 | Conclusion & More Resources

- Apple Music for Beginners
- Online Safety: The Complete Guide to Being Safe Online
- Smartphones & Tablets for Beginners
- Instagram App for Beginners
- Snapchat App for Beginners
- iTunes for Beginners
- Mac Computer for Beginners
- Making your Computer Fast Again Tutorial
- Kindle Manual for Beginners
- Fire HD Tablet Manual for Beginners

More Resources – The Digitize Center

Your iPhone is an excellent place to store all your photos and videos tapes, even ones from the past that are sitting in old photo albums or boxes in your closet. You can easily get all your old memories onto your iPhone by using The Digitize Center, a company that digitizes old media types such as photos, slides, film reels, VHS tapes, camcorder tapes and more. Once your items are digitized, The Digitize Center can send them directly to your iPhone where you can easily access and share them.

I recommend using The Digitize Center if you are ever in need of getting your precious memories digitized, and they make it extremely easy to get your memories on to your iPhone. You can learn more about this on their website at www.thedigitizecenter.com.

Use the code APPLE for 10% OFF.

The Digitize Center
PRESERVING PRECIOUS MEMORIES

Chapter 19 | Conclusion & More Resources

Check out more beginner's guides and manuals at:

www.infinityguides.com

Notes

Index

A

Action Button 16
Activity & Mobility Data 83
Activity App 65
Airplane Mode 51
AirPlay 53
AirPods 53
App Switcher 101
Apple Pay 95
Apple Watch
    about 13
    charging 17
    layout 15
    Models *14*
    turning on & off 16
    waking 17
Apps 94
    definition *13*
Apps Screen 106

B

Background Apps 101
Bands 13
battery 102
Battery 51
Breathe 70

C

Camera 103
Cellular 14, 50
Compass App 91
Complications 41
Control Center 49
Crash Detection 88
Cycle Tracking 80

D

Digital Crown 15, 29
Do Not Disturb 51
Double Tap 103

E

ECG App 73
Emergency Services 86
Exercise 67

F

Face Gallery 35
Faces 31, 33
Fall Detection 87
Find My iPhone 51
First-time Setup 19
Fitness 67
Flashlight 51

G

Gesture 103

H

Health App 81
Health Features 69
Health Summary 81
Heart App 75
Heart Data 82
Home Face
    definition *13*

I

Infinity Guides *107*
iPhone 13
iPhone Manual for Beginners 107

111

Index

L

Low Power Mode 102

M

Mail App 89
Medical ID 85
Mindfulness App 69
Music App 92
My Watch 34

N

Navigating 29
Noise App 78
Notification Center 55
Notifications 55
Now Playing App 92

P

Pairing your Apple Watch 19
Phone Calls 61

R

Reflect 70

S

Safety Features 85
side button 16
Side Button 31

silent mode 51
Siren 88
Siri 99
Sleep App 70
Sleep Data 83
Smart Stack 102
State of Mind 69
Stopwatch App 90

T

Text Messaging 57
Text Size 52
The Digitize Center 108
Theater Mode 51
Timer App 89

V

Vitals App 81
volume 52

W

Walkie-Talkie 53
Watch App 33
watchOS 14, 27
water lock 51
Weather App 91
Wi-Fi 50
Workout App 67